Photonic Biosensors: Detection, Analysis and Medical Diagnostics

Photonic Biosensors: Detection, Analysis and Medical Diagnostics

Editor

Donato Conteduca

MDPI • Basel • Beijing • Wuhan • Barcelona • Belgrade • Manchester • Tokyo • Cluj • Tianjin

Editor
Donato Conteduca
University of York
UK

Editorial Office
MDPI
St. Alban-Anlage 66
4052 Basel, Switzerland

This is a reprint of articles from the Special Issue published online in the open access journal *Biosensors* (ISSN 2079-6374) (available at: https://www.mdpi.com/journal/biosensors/special_issues/Photonic_Biosensors_DAD).

For citation purposes, cite each article independently as indicated on the article page online and as indicated below:

LastName, A.A.; LastName, B.B.; LastName, C.C. Article Title. *Journal Name* **Year**, *Volume Number*, Page Range.

ISBN 978-3-0365-3971-3 (Hbk)
ISBN 978-3-0365-3972-0 (PDF)

Contents

About the Editor

Donato Conteduca received his MSc degree (2013) in Optoelectronics from Politecnico di Bari (Italy) and completed an internship with the Microphotonics and Photonic Crystals Group at the University of St. Andrews (Scotland). He completed his PhD at Politecnico di Bari in 2017, and the project involved the realization of an optofluidic system for optical trapping of sub-micrometer particles in collaboration with the Photonics Group at the University of York. He is currently a Postdoctoral Researcher in the Photonics Group at the University of York. His Research focuses on integrated optical nanotweezers for trapping at nanoscale and nanophotonic devices for biochemical sensing that can be translated for use in clinical applications. Donato Conteduca has published over 50 scientific papers in international journals and has served as guest editor for journals published by MDPI and Frontiers.

 *biosensors*

Editorial

Photonic Biosensors: Detection, Analysis and Medical Diagnostics

Donato Conteduca

Photonics Group, University of York, York YO10 5DD, UK; donato.conteduca@york.ac.uk

Citation: Conteduca, D. Photonic Biosensors: Detection, Analysis and Medical Diagnostics. *Biosensors* **2022**, *12*, 238. https://doi.org/10.3390/bios12040238

Received: 8 April 2022
Accepted: 12 April 2022
Published: 13 April 2022

The necessity of personalised diagnoses and ad hoc treatments for individual patients is driving the outbreak of personalised nanomedicine in research and in clinical studies in the healthcare field. In this context, nanotechnologies are assuming a fundamental role due their ability to promptly diagnose a disease, its progression and aggressiveness, and their capability to define the most efficient drugs and treatments due to their continuous screening ability, making treatments less and less harmful for patients [1].

Together with the strong efforts to improve sensor performance, the demand to develop biosensors in low-cost and user-friendly technologies can no longer be neglected due to a forthcoming use of these sensing platforms in hospitals and clinics by healthcare workers.

Photonics biosensors have demonstrated their ability to combine high sensitivity, selectivity, and reliability together with real-time operation and strong integrability in cost-efficient solutions. Furthermore, the capability to integrate photonic biosensors with microfluidics and the compatibility of most of the photonic architectures and materials with electric readouts make them the most suitable candidates for the development of lab-on-a-chip systems to be integrated in point-of-care instruments [2].

The main goal of this Special Issue has been to address new technologies, materials, and configurations for photonic biosensors that can improve the sensitivity, accuracy, and precision of medical diagnostics. Novel biomedical studies have also been highlighted in this collection, i.e., for cardiovascular diseases and bacterial infections, which are now possible due to the technological progress and innovation that have been made in this field. Moreover, the strong demand for the integrability of the photonic biosensors in compact systems, which is playing a more fundamental role in many sensing applications, such as in wearable and implanted sensors, is becoming more evident.

The published articles, which are briefly described in the following section, cover these topics and make a strong impact in the biosensing field.

In [3], Adamopoulos et al. describe a novel strategy to create both micro and meso-scale microfluidic structures that were developed on multiple levels for integrated into monolithic photonic–electronic systems on a single chip. The strong advantage of silicon photonic sensors is their strong integrability with a printed circuit board (PCB) for the electrical readout and their compactness, allowing a large number of devices in a single photonic chip to undergo advanced multiplexing. However, the drawback is mainly related to the interface of the photonic sensors with the standard microfluidics. Standard packaging techniques usually suffer of this extreme compactness, which does not allow each sensor to be accessed and addressed without affecting the final cost of the chip. The authors solve this typical issue by creating a small microfluid interface that bonds to the silicon chip, accesses the single device, and uses micro-ring resonators as a sensing element as proof of concept. This microfluidic interfaces with another layer through vertical connections, where macro-scale tubes are directly connected. The realization of the multilevel microfluidics through the adoption of 3D printed transfer moulding and bonding can guarantee high scalability and cost efficiency.

In [4], Saha et al. developed a wearable device to measure blood circulation in human patients. The possibility of integrating the whole sensor in a watch-sized device represents a key technology for the next generation of biosensors because it allows for the continuous monitoring of health parameters in a fully non-invasive manner, something that could impact personalised medicine.

The authors developed a sensor based on laser Doppler flowmetry. The external sensor probe includes both an optical source and detector for the real-time measurement of blood perfusion by analysing the Doppler spectrum. The pilot study reported in this paper confirms the possibility of differentiating the cardiovascular parameters between smokers and non-smokers, findings that were confirmed by lower blood perfusion in the latter group and the presence of an embedded accelerometer that allows the blood perfusion levels to be discriminated from actual body movements.

A non-invasive sensing modality based on near-infrared (NIR) spectroscopy was also used by Heise et al. in [5] to measure blood glucose levels. The science-based calibration (SBC) method was applied by the authors and, in particular, tissue glucose concentrations were measured using the lip NIR-spectra of a type-1 diabetic patient recorded under modified oral glucose tolerance test (OGTT) conditions.

The reflection spectroscopy technique is preferred to transmission measurements because the latter requires thin skin folds or short-wave NIR spectroscopy to illuminate a fingertip or an earlobe. The main achievements of the above work are the demonstration that SBC calibration can be adopted for non-invasive studies and when selective measurements are needed. As confirmed by the authors, for further developments in this technology, more model parameters will be considered for in vivo spectroscopy, with physiology, sensor repositioning, temperature gradients, and blood flow changes being only a few of the main variables that are to be investigated in detail.

The necessity of continuous blood pressure screening using portable instruments and cost-efficient solutions is a recurring and pressing topic in healthcare technologies. Sun et al. in [6] adopted a convolutional neural network (CNN) based on the Hilbert–Huang Transform method, to predict the blood pressure risk level using photoplethysmography (PPG). The authors also proved that the derivatives of PPG carry important information on blood pressure, which could allow the combination of the ECG and PPG techniques to be replaced. Deep learning will be investigated in more detail in further studies in order to improve the accuracy of the measurements and the prediction of the blood pressure values related to specific physiological activities.

Rho et al. in [7] propose an optical cavity-based biosensor used for streptavidin and C-reactive protein (CRP) sensing. The optical cavity consists in two partially reflective and parallel surfaces separated by a thin gap, which affects the properties of the optical spectrum. Therefore, the gap is considered to be the sensing area that is functionalised with specific receptors. When the target analyte binds to the receptor, the changes in the optical properties provide changes in the optical intensities at two different wavelengths. The multi-wavelength system allows the cost of the system to be minimized by replacing the tunable laser and the spectrometer with low-cost laser diodes and a CMOS camera. The sensing modality is based on a differential detection method that maximizes the sensitivity, dynamic range, and fabrication tolerances. The authors have verified a limit-of-detection (LOD) of 377 pM for CRP by using a small sample volume of only 15 μL within 30 min.

Finally, in [8], Brunetti et al. modelled an optoelectronic biosensor to monitor bacterial biofilm evolution based on a multiparameter analysis with optical and electrical detection schemes. The sensor consists of a dual array comprising interdigitated micro- and nano-electrodes in parallel. The optical response of the nanostructure is typical of a 1D grating supporting guided mode resonance (GMR). The resonant response allows the imaging of planktonic bacteria distributed in small colonies to be carried out at a higher spatial resolution, providing useful information on the modality of biofilm generation. In addition, the electrical response of both micro- and nanoelectrodes is necessary for the study of the metabolic state of the bacteria, which can be very powerful in the testing of the efficacy of antibiotics on the biofilm. The main advantage of this sensing configuration based on a multiparameter approach is the capability to detect and monitor a biofilm in real-time while also analysing its metabolic state and the evolution phase. Optoelectronic devices can make a strong impact in antimicrobial resistance studies.

Acknowledgments: D.C. was supported by the Engineering and Physical Sciences Research Council (EPSRC) of the UK (Grant EP/P030017/1).

Conflicts of Interest: The authors declare no conflict of interest.

References

1. Conteduca, V.; Brighi, N.; Conteduca, D.; Bleve, S.; Gianni, C.; Schepisi, G.; Iaia, M.L.; Gurioli, G.; Lolli, C.; de Giorgi, U. An update on our ability to monitor castration-resistant prostate cancer dynamics with cell-free DNA. *Expert Rev. Mol. Diagn.* **2021**, *21*, 631–640. [CrossRef] [PubMed]
2. Min, J.; Son, T.; Hong, J.; Cheah, P.S.; Wegemann, A.; Murlidharan, K.; Weissleder, R.; Lee, H.; Im, H. Plasmon-Enhanced Biosensing for Multiplexed Profiling of Extracellular Vesicles. *Adv. Biosyst.* **2020**, *4*, 2000003. [CrossRef] [PubMed]
3. Adamopoulos, C.; Gharia, A.; Niknejad, A.; Stojanovic, V.; Anwar, M. Microfluidic Packaging Integration with Electronic-Photonic Biosensors Using 3D Printed Transfer Molding. *Biosensors* **2020**, *10*, 177. [CrossRef] [PubMed]
4. Saha, M.; Dremin, V.; Rafailov, I.; Dunaev, A.; Sokolovski, S.; Rafailov, E. Wearable Laser Doppler Flowmetry Sensor: A Feasibility Study with Smoker and Non-Smoker Volunteers. *Biosensors* **2020**, *10*, 201. [CrossRef] [PubMed]
5. Heise, M.; Delbeck, S.; Marbach, R. Noninvasive Monitoring of Glucose Using Near-Infrared Reflection Spectroscopy of Skin—Constraints and Effective Novel Strategy in Multivariate Calibration. *Biosensors* **2021**, *11*, 64. [CrossRef] [PubMed]
6. Sun, X.; Zhou, L.; Chang, S.; Liu, Z. Using CNN and HHT to Predict Blood Pressure Level Based on Photoplethysmography and Its Derivatives. *Biosensors* **2021**, *11*, 120. [CrossRef] [PubMed]
7. Rho, D.; Kim, S. Demonstration of a Label-Free and Low-Cost Optical Cavity-Based Biosensor Using Streptavidin and C-Reactive Protein. *Biosensors* **2021**, *11*, 4. [CrossRef] [PubMed]
8. Brunetti, G.; Conteduca, D.; Armenise, M.N.; Ciminelli, C. Novel Micro-Nano Optoelectronic Biosensor for Label-Free Real-Time Biofilm Monitoring. *Biosensors* **2021**, *11*, 361. [CrossRef] [PubMed]

 *biosensors*

Communication

Microfluidic Packaging Integration with Electronic-Photonic Biosensors Using 3D Printed Transfer Molding

Christos Adamopoulos [1,*,†], Asmaysinh Gharia [1,2,†], Ali Niknejad [1], Vladimir Stojanović [1] and Mekhail Anwar [2]

[1] Department of Electrical Engineering and Computer Science, University of California Berkeley, Berkeley, CA 94720, USA; agharia@berkeley.edu (A.G.); niknejad@berkeley.edu (A.N.); vlada@berkeley.edu (V.S.)
[2] Department of Radiation Oncology, University of California San Francisco, San Francisco, CA 94158, USA; Mekhail.Anwar@ucsf.edu
*   Correspondence: christos.ad@berkeley.edu
†   These authors contributed equally to this work.

Received: 14 October 2020; Accepted: 12 November 2020; Published: 14 November 2020

Abstract: Multiplexed sensing in integrated silicon electronic-photonic platforms requires microfluidics with both high density micro-scale channels and meso-scale features to accommodate for optical, electrical, and fluidic coupling in small, millimeter-scale areas. Three-dimensional (3D) printed transfer molding offers a facile and rapid method to create both micro and meso-scale features in complex multilayer microfluidics in order to integrate with monolithic electronic-photonic system-on-chips with multiplexed rows of 5 μm radius micro-ring resonators (MRRs), allowing for simultaneous optical, electrical, and microfluidic coupling on chip. Here, we demonstrate this microfluidic packaging strategy on an integrated silicon photonic biosensor, setting the basis for highly multiplexed molecular sensing on-chip.

Keywords: integrated photonics; microfluidics; packaging; photonic biosensors; optical resonators; multiplexed sensing; 3D printing

1. Introduction

Label-free assays provide real time information for molecular analysis without requiring labeling of the target analytes that require customized molecular labels, multiple additional steps, and preclude monitoring real-time binding kinetics useful for characterizing molecular interactions. This makes them ideal for point-of-care assays. However, to date, label-free assays have been hindered by the need for complex, bulky optics precluding both mass production and miniaturization and remain confined to special laboratory settings [1–3]. In an effort to address these dual challenges, silicon photonic biosensors—fabricated using techniques based on integrated circuit manufacturing—have shown increasing promise in monitoring label-free molecular interactions through the evanescent field interaction of light, tightly confined to a waveguide, with the surrounding optical environment. In particular, one such structure where light of a specific wavelength is trapped in a ring (called a micro-ring resonator or micro-ring resonator (MRR)-based photonic devices), has been able to detect a wide range of molecular concentrations of different target analytes ranging from femtomolar to micromolar by leveraging high transducer sensitivity and small transducer area [4–13]. However, one of the key challenges towards efficient and diverse multiplexed sensing of multiple biomarkers is the fabrication of fluidic delivery devices to interface with the extraordinary density and degree of integration of arrays of micrometer-scale MRRs, while allowing for both optical and electrical coupling to the integrated circuit, which is often on a millimeter-square scale. Microfluidic interfaces with silicon photonic chips require (a) regular surfaces to form a seal and prevent leaking, (b) optical clarity for alignment, (c) vertical cylindrical interconnections with an aspect ratio of 5, which is defined as the ratio of their height to their width, to connect microfluidic layers with a high channel density (100 µm–200 µm channel width and edge to edge channel spacing), (d) access areas allowing fiber optics to couple light into the system, and (e) a mechanism to interface multiple microfluidic inputs to millimeter-scale chips with densely packed sensor arrays.

The current approaches to meet these constraints often require custom packaging techniques in order to accommodate centimeter scale microfluidics [14] or complex post processing following conventional multilayer soft lithography [15]. The bonding of microfluidics to chips typically requires a broad, flat surface in order to provide an adequate seal and structural support for the microfluidics. A key challenge arises with the small chip size (square millimeter-scale), constrained by the cost of manufacturing large (square centimeter-scale) chips. Large chip surfaces cannot be manufactured due to the high cost per unit area of manufacturing silicon-based devices. Thus, many small sensor elements are packed together in tightly spaced arrays. This makes it conventionally challenging to interface with standard microfluidics and inlet-outlet tubes, which are on the order of 5 mm diameter (needed to interface with standard pumps and sample sources), since they do not fit on the area of a traditional chip. While there are many techniques to create a flat surface for attaching microfluidic devices [14], we require a method to interface millimeter scale microfluidics with a small chip. We create a small interface microfluidic chip that securely bonds on the silicon photonic chip [15,16]. By doing this, we eliminate the need for larger chip areas that would significantly increase the cost. This bottom microfluidic layer then interfaces through vertical interconnections with another layer that expands microfluidic dimensions, so that they can interface with standard sized microfluidics and macro-scale tubing over a larger surface area.

This multilayer interface approach requires precisely aligned microfluidics at two different scales. Creating each layer via photolithography traditionally offers the greatest precision in channel dimensions and surface finishes, but it requires complex processing to create high-density vertical interconnections and align multiple layers [17]. A hybrid of soft lithography and laser micromachining can be used to address these challenges but requires expensive and specialized equipment [15]. Additive manufacturing can greatly reduce the complexity in fabricating such multilayer devices through 3D printed transfer molding wherein each layer is cast on a printed mold. Despite the promise of 3D printing, the fabrication of molds for microfluidics has been historically limited by the inability to produce the smooth surface finishes—necessary for both visualization through the microfluidic layers as well as proper sealing—and minimum feature sizes of photolithography, and has thus been limited to centimeter scale microfluidics [18].

Commercial printing technology has now crossed the resolution threshold in order to overcome surface finish, feature size, and aspect ratio constraints in designing millimeter scale microfluidics. Leveraging these advances, we demonstrate a simple, highly scalable, and versatile multilayer microfluidic device for CMOS integration created by 3D printed transfer molding and bonding. This packaging architecture can enable multiplexed analyte sensing without requiring large chip surfaces and complicated fabrication techniques, while it ensures a robust interface between the photonic sensors and the fluidic samples.

2. Materials and Methods

Three-dimensional (3D) printed transfer molding enables rapid fabrication of multilayer microfluidics for co-packaging with highly miniaturized silicon photonic chips, as seen in Figure 1.

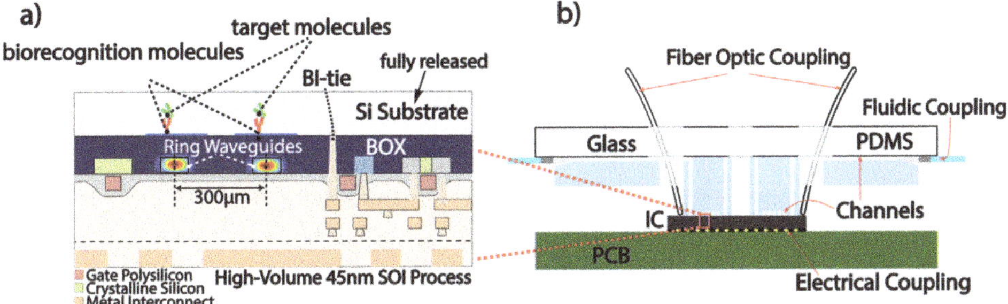

Figure 1. (**a**) CMOS 45RFSOI process cross-section. Backside Si substrate etch exposes the sensing photonics to the fluidic samples. (**b**) Cross section of microfluidic package aligned to chip. Photonic ports allow for fiber optic coupling, needles inserted at the edge allow for fluidic coupling, and flip chip bonding allows for electrical coupling. Layer 1 interfaces with the chip, layer 2 routes fluid to tubing, and layer 3 provides mechanical support.

This versatile packaging technique can accommodate multi-fluidic and photonic coupling, independent of the electronic interface. As a first step, the advanced process node of the 45 nm Silicon on Insulator (SOI) fully integrated electronic-photonic platform requires removal of the Si substrate to prevent leakage of the optical mode in the substrate and expose the photonic sensors to the fluidic samples, as illustrated in Figure 1a. The Si substrate of the silicon-photonic chip is fully removed via a XeF_2 etch. Because XeF_2 is highly selective to silicon over silicon dioxide (SiO_2), the buried oxide layer (BOX), which is SiO_2-based, acts as a natural etch stop. The XeF_2 etch is performed in excess until the silicon is visually observed to be removed. In the package that is shown in Figure 1b, the silicon chip is flip chip bonded on a printed circuit board (PCB) for electrical coupling and multilayer microfluidics are mounted on top of the chip. Our microfluidic design consists of two polydimethylsiloxane (PDMS) layers and a glass substrate for mechanical stability. The layers are enumerated by proximity to the chip, wherein the small area of the PDMS layer in contact with the chip is layer 1 ("primary layer"), the larger layer for fluid routing is layer 2 ("secondary layer"), and finally the glass substrate is layer 3.

2.1. 3D Printed Mold Fabrication

The mold for each PDMS layer is designed in CAD (Autodesk Fusion360) as an array of devices with various geometries of vertical interconnections (vias), channels, and access areas for optic fibers that act as photonic ports. We designed devices that range from four to 14 channels, eight to 28 vias, and three photonic ports on a 5.5 mm by 3 mm area to match the size of our sensing chip in order to test the limitations of the printed molds. The final 3D model forms interconnections with a height to width ratio between 5:1 to 10:1, high resolution channels, meso-scale photonic ports, and alignment marks as demonstrated in Figure 2. Alignment marks aid in precisely locating multiple layers while assembling the final device.

The model was submitted for fabrication by projection micro-stereolithography (Boston Micro Fabrication) with 2 µm resolution and using UV curable acrylate-based resins as the material. Bottom layer molds with varying channel widths were fabricated, achieving a smooth surface finish with interconnections of 5:1 to 10:1 height to width ratio, millimeter scale photonic ports, and an edge to edge channel spacing ranging from 150 µm to 200 µm, thus meeting the requirements for on-chip silicon photonic integration. The channel pitch was fixed at 300 µm set by the micro-ring resonator pitch, and the channel width was varied to assess risk of channel to channel leakage. Earlier attempts of printing similar molds while using high resolution fused deposition modeling and stereolithography printers failed to produce the necessary surface finish. Meanwhile, extremely high resolution 2 photon process printers could not produce larger meso-scale features.

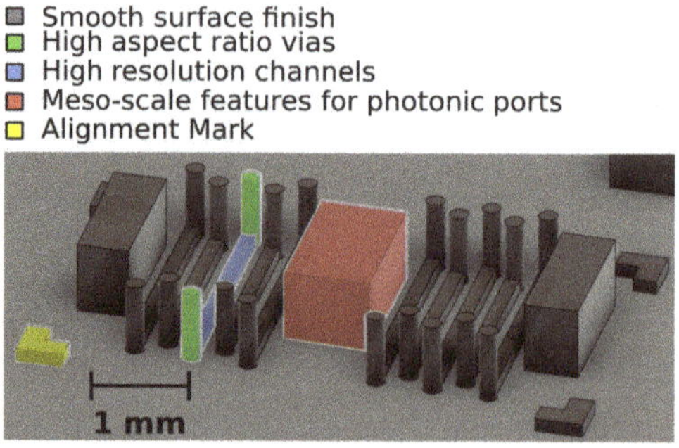

Smooth surface finish
High aspect ratio vias
High resolution channels
Meso-scale features for photonic ports
Alignment Mark

1 mm

Figure 2. 3D printed mold model for 10 channel device with 20 via-interconnections, 3 photonic ports, and alignment marks.

2.2. Microfluidic Device Casting

The mold must be coated in a release agent prior to casting for a high-fidelity transfer of features without damage to delicate components of both the mold and the cast. To achieve this, the mold (Figure 3a) is placed in a vacuum chamber with 500 µL of Trichloro(1H, 1H, 2H, 2H-perfluorooctyl)silane (Sigma) under vacuum for 15 min. [19], as shown in Figure 3b. PDMS (Sylgard 184, Dow) is a composite of two materials: a pre-polymer, or material that can form into a polymer, and a crosslinking agent, which, when mixed with the pre-polymer, connects the molecules in a network that forms an elastic substrate. The pre-polymer and crosslinker are thoroughly mixed at the manufacturer's recommended 10 parts pre-polymer to one part crosslinker by weight and degassed under vacuum (15 mbar) for 45 min. [20]. This mixture is poured onto the silanized mold (Figure 3c) and it cures into an elastic substrate that is patterned with the mold's features. The mold is capped with a cleaned glass slide (Figure 3d) to ensure an even top surface with smooth finish required for multilayer bonding.

A 500-gram weight is then placed on top of the glass to remove residual PDMS between contact areas on the mold and glass in order assure thru-hole formation for vertical interconnections and photonic ports rather than thin membranes. The mold, PDMS, and glass sandwich are placed on a hot-plate at 95 degrees Celsius for 2 h to cure the elastomer. After the PDMS has cured, the cast device is removed from the device and then trimmed with a razor to its final size (Figure 3e). This process is repeated for each layer of the final multilayer device.

2.3. Alignment and Mechanical Sealing

Once layers 1 and 2 have been cast, a glass substrate with milled photonic ports (layer 3) is prepared in order to provide mechanical support to the microfluidic channel networks. Layer 2, denoted as upper PDMS in Figure 4a, is first bonded to layer 3 with oxygen plasma by placing them in a reactive ion etch chamber with the mating surfaces exposed. Plasma introduces reactive hydroxyl groups on the surface that facilitate irreversible chemical bonding [21]. The mating surfaces are then aligned under magnification to maximize overlap between the photonic ports of the glass substrate and PDMS layer.

This assembly is placed on a hot plate at 65 degrees Celsius for 15 minutes to drive the reaction to completion. This process is repeated for bonding layer 1 to the layers 2,3 assembly. Coarse alignment is achieved by aligning corner holes in the glass substrate (Figure 4b) with holes on the PCB with screws. Fine alignment is further manually achieved by leveraging optical transparency through layers of the microfluidic stack for visual alignment of channels to sensing structures on-chip. Once aligned, the channels are sealed to the chip through mechanical pressure by tightening corner screws. Bonding the microfluidics to the chip exclusively through mechanical pressure allows for thorough decontamination of the chip surface between tests by entirely removing the package for reuse of the electronic-photonic chip. It also enables the rapid iteration of various channel geometries without requiring a unique chip to test each geometry.

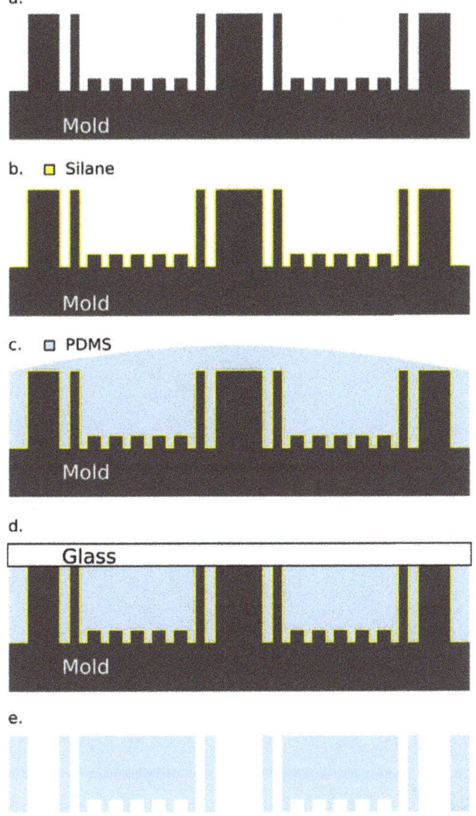

Figure 3. Workflow for casting polydimethylsiloxane (PDMS) on the three-dimensional (3D) printed mold. (**a**) The mold is (**b**) silanized in a vacuum chamber. (**c**) Mixed PDMS is poured onto the mold and (**d**) the top surface is made uniform with a glass cap. (**e**) The PDMS is then cured and removed from the mold.

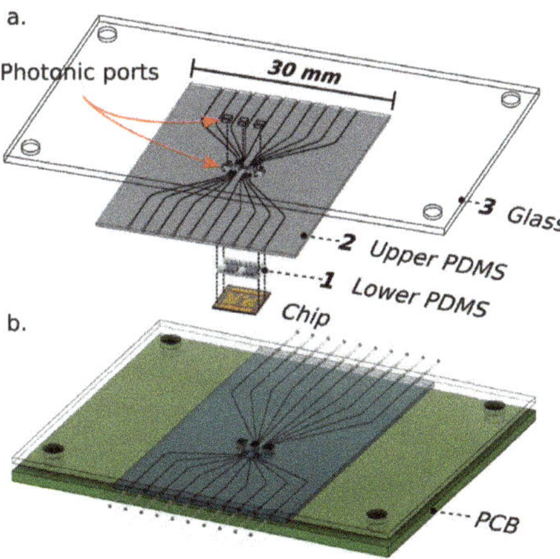

Figure 4. (**a**) Multilayer stack of glass substrate and cast PDMS pieces are aligned and then bonded using oxygen plasma. (**b**) The assembly is then coarsely aligned to the printed circuit board (PCB) before fine alignment of channels to sensing regions.

3. Results and Discussions

3.1. Microfluidic Validation—Dye Test

We demonstrate the functionality of the described microfluidic packaging by injecting colored dyes to detect leaks between channels. In Figure 5, colored dyes are injected into a 10-channel microfluidic device with 150 μm edge to edge channel spacing using edge coupling tubing with 28-gauge needles. One channel is kept empty as a reference. There was no cross contamination of dye in either the lower chip-sized PDMS channels or the larger upper PDMS channels. This test was done on glass for high contrast to best visualize any leaks. Once the proof of concept was verified while using dye on glass, a 10-channel microfluidic package with a 200 μm edge to edge channel spacing was aligned to 5 μm radius MRRs and optic fibers were aligned with grating couplers on-chip for photonic coupling.

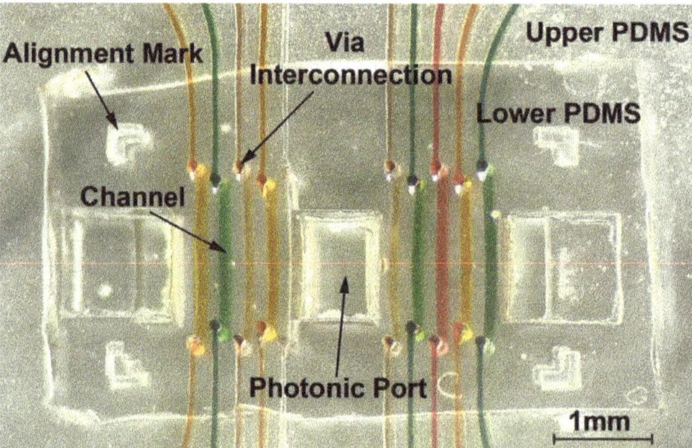

Figure 5. Colored dyes run through the multi-layer microfluidic assembly.

3.2. Photonic Coupling and Bulk Sensitivity

Channels of the assembly were aligned to rings, as described above. Subsequently, 125 μm diameter lensed fibers are lowered through the photonic ports while using micromanipulators on an optical table. These fibers are aligned to on-chip grating couplers to couple laser light into a waveguide. When the wavelength of the input light is at the resonant wavelength of an MRR, power couples into, and circulates within, the micro-ring resulting in a power loss at the output coupler [22]. Figure 6 is an infrared image of channels that are aligned to the MRRs of a fully integrated electronic-photonic system on-chip. Photonic coupling is demonstrated by the circulating power emanating from the micro-rings.

The bulk sensing capability of the platform was then evaluated using the proposed microfluidic packaging strategy. Light from a tunable laser (SANTEC TSL-510), which has the ability to change the wavelength of light output under fine control, is guided to a row of ring resonators. For these experiments, the light from the laser was swept from 1298 nm to 1308 nm. The laser performs multiple wavelength sweeps of the ring row's output at a step of 10 picometers (pm). The output power is measured with a power meter (Agilent 8164B). For each sweep, the resonant wavelengths of the rings are captured and their difference from the nominal resonances of the initial sweep (resonant shift) is calculated. Figure 7 shows the resonant shift of three rings that were located in three different channels. Channels 1 and 2 are exposed to water, while channel 3 is empty. A resonant shift of 650pm was measured for the sensors under water. By characterizing the RI of water with an ATC refractometer (RI = 1.3334), a bulk sensitivity of 1.95 nm/Refractive Index Unit is calculated by dividing the resonant shift with the change of the RI from air to water ($\Delta RI = 0.3334$). This high channel density and miniaturized microfluidic device has the potential to integrate with a variety of silicon photonic devices.

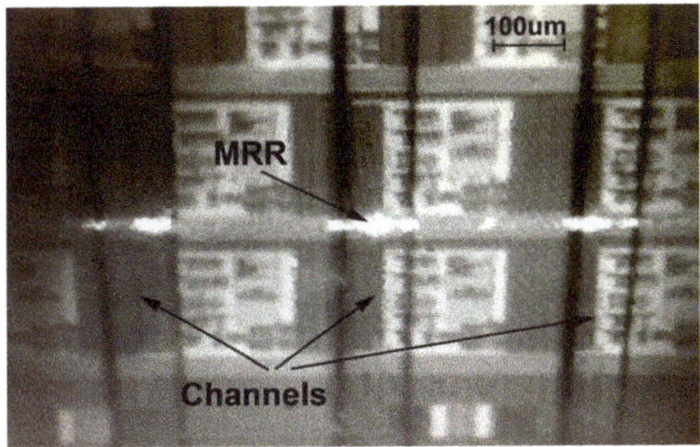

Figure 6. Power loss through micro-ring resonators while aligned to channels. This close up focuses on 3 channels from the 10-channel package of Figure 4.

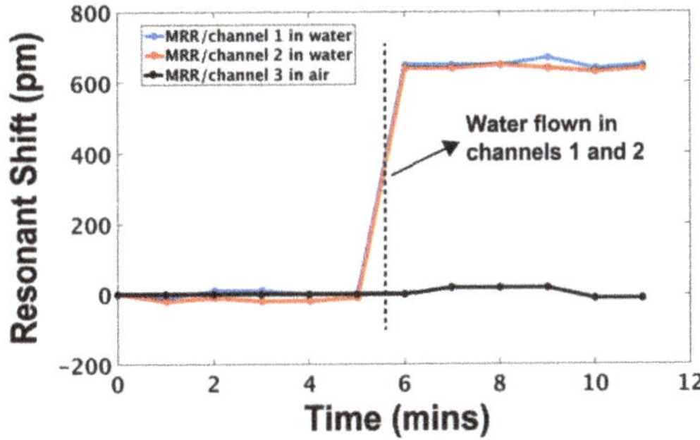

Figure 7. Resonant shift of micro-ring resonators (MRRs) exposed to water and air.

4. Conclusions

Commercial projection micro stereolithography with sub 2 μm resolution paves the way for using three-dimensional (3D) printed transfer molding to create silicon photonic compatible microfluidics. The simultaneous success of fluidic, photonic, and electronic coupling using this rapid packaging strategy underlies its versatility. The facile fabrication of multilayer dense channel networks with smooth surface finish and photonic ports can solve key challenges for multi-fluid coupling, alignment of channels to micron scale sensing elements, and photonic coupling to the device. This unlocks the door towards compact and self-contained biophotonic sensors.

Funding: This research is supported in part by the National Institutes of Health 3U54HL119893-07S1. Mekhail Anwar is also supported by the National Institutes of Health DP2DE030713, R21EB027238.

Acknowledgments: We would like to thank the National Institutes of Health, the Berkeley Wireless Research Center (BWRC), Santec, Boston Micro Fabrication and the Biomolecular Nanotechnology Center (BNC) at UC Berkeley for support.

Conflicts of Interest: The authors declare no conflict of interest.

References

1. Yang, C.-Y.; Brooks, E.; Li, Y.; Denny, P.; Ho, C.-M.; Qi, F.; Shi, W.; Wolinsky, L.; Wu, B.; Wong, D.T.W.; et al. Detection of picomolar levels of interleukin-8 in human saliva by SPR. *Lab Chip* **2005**, *5*, 1017–1023. [CrossRef] [PubMed]
2. Jason-Moller, L.; Murphy, M.; Bruno, J. Overview of Biacore systems and their applications. *Curr. Protoc. Protein Sci.* **2006**, *45*, 19.13.1–19.13.14. [CrossRef] [PubMed]
3. Gopinath, S. Biosensing applications of surface plasmon resonance-based Biacore technology. *Sens. Actuators B Chem.* **2010**, *150*, 722–733. [CrossRef]

4.   Washburn, A.L.; Luchansky, M.S.; Bowman, A.L.; Bailey, R.C. Quantitative, Label-Free Detection of Five Protein Biomarkers Using Multiplexed Arrays of Silicon Photonic Microring Resonators. *Anal. Chem.* **2010**, *82*, 69–72. [CrossRef] [PubMed]

5.   Washburn, A.L.; Gunn, L.C.; Bailey, R.C. Label-free quantitation of a cancer biomarker in complex media using silicon photonic microring resonators. *Anal. Chem.* **2009**, *81*, 9499–9506. [CrossRef]

6.   Luchansky, M.S.; Washburn, A.L.; McClellan, M.S.; Bailey, R.C. Sensitive on-chip detection of a protein biomarker in human serum and plasma over an extended dynamic range using silicon photonic microring resonators and sub-micron beads. *Lab Chip* **2011**, *11*, 2042–2044. [CrossRef]

7.   Qavi, A.J.; Kindt, J.T.; Gleeson, M.A.; Bailey, R.C. Anti-DNA:RNA Antibodies and Silicon Photonic Microring Resonators: Increased Sensitivity for Multiplexed microRNA Detection. *Anal. Chem.* **2011**, *83*, 5949–5956. [CrossRef]

8.   Iqbal, M.; Gleeson, M.A.; Spaugh, B.; Tybor, F.; Gunn, W.G.; Hochberg, M.; Baehr-Jones, T.; Bailey, R.C.; Gunn, L.C. Label-Free Biosensor Arrays Based on Silicon Ring Resonators and High-Speed Optical Scanning Instrumentation. *IEEE J. Sel. Top. Quantum Electron.* **2010**, *16*, 654–661. [CrossRef]

9.   Flueckiger, J.; Schmidt, S.; Donzella, V.; Sherwali, A.; Ratner, D.M.; Chrostowski, L.; Cheung, K.C. Sub-wavelength grating for enhanced ring resonator biosensor. *Opt. Express* **2016**, *24*, 15672. [CrossRef]

10.   Yan, H.; Huang, L.; Xu, X.; Chakravarty, S.; Tang, N.; Tian, H.; Chen, R.T. Unique surface sensing property and enhanced sensitivity in microring resonator biosensors based on subwavelength grating wave-guides. *Opt. Express.* **2016**, *24*, 29724–29733. [CrossRef]

11.   Adamopoulos, C.; Gharia, A.; Niknejad, A.; Anwar, M.; Stojanović, V. Electronic-Photonic Platform for Label-Free Biophotonic Sensing in Advanced Zero-Change CMOS-SOI Process. In Proceedings of the Conference on Lasers and Electro-Optics, OSA, San Jose, CA, USA, 5–10 May 2019; p. JW2A.81. [CrossRef]

12.   Claes, T.; Molera, J.G.; De Vos, K.; Schacht, E.; Baets, R.; Bienstman, P. Label-Free Biosensing With a Slot-Waveguide-Based Ring Resonator in Silicon on Insulator. *IEEE Photonics J.* **2009**, *1*, 197–204. [CrossRef]

13.   Vos, K.D.; Girones, J.; Claes, T.; Koninck, Y.D.; Popelka, S.; Schacht, E.; Baets, R.; Bienstman, P. Multiplexed Antibody Detection With an Array of Silicon-on-Insulator Microring Resonators. *IEEE Photonics J.* **2009**, *1*, 225–235. [CrossRef]

14.   Khan, S.; Gumus, A.; Nassar, J.; Hussain, M. CMOS Enabled Microfluidic Systems for Healthcare Based Applications. *Adv. Mater.* **2018**, *30*, 1705759. [CrossRef] [PubMed]

15.   Muluneh, M.; Issadore, D. A multi-scale PDMS Fabrication strategy to bridge the size mismatch between integrated circuits and microfluidics. *Lab Chip* **2014**, *14*, 4552–4558. [CrossRef] [PubMed]

16.   Wu, H.; Odom, T.; Chiu, D.; Whitesides, G. Fabrication of complex three-dimensional microchannel systems in PDMS. *J. Am. Chem. Soc.* **2003**, *125*, 554–559. [CrossRef] [PubMed]

17.   Comina, G.; Suska, A.; Filippini, D. PDMS lab-on-a-chip fabrication using 3D printed templates. *Lab Chip* **2014**, *14*, 424–430. [CrossRef] [PubMed]

18.   Qin, D.; Xia, Y.; Whitesides, G. Soft lithography for micro- and nanoscale patterning. *Nat. Protoc.* **2010**, *5*, 491–502. [CrossRef]

19.   Zhang, M.; Wu, J.; Wang, L.; Xiao, K.; Wen, W. A simple method for fabricating multi-layer PDMS structures for 3D microfluidic chips. *Lab Chip* **2010**, *10*, 1199. [CrossRef]

20.   Friend, J.; Yeo, L. Fabrication of microfluidic devices using polydimethylsiloxane. *Biomicrofluidics* **2010**, *4*, 026502. [CrossRef]

21.   Bhattacharya, S.; Datta, A.; Berg, J.M.; Gangopadhyay, S. Studies on surface wettability of poly(dimethyl) siloxane (PDMS) and glass under oxygen-plasma treatment and correlation with bond strength. *J. Microelectromech. Syst.* **2005**, *14*, 590–597. [CrossRef]

22.   Bogaerts, W.; de Heyn, P.; van Vaerenbergh, T.; de Vos, K.; Selvaraja, S.K.; Claes, T.; Dumon, P.; Bienstman, P.; van Thourhout, D.; Baets, R. Silicon microring resonators. *Las. Photon.* **2012**, *6*, 590- [CrossRef]

Publisher's Note: MDPI stays neutral with regard to jurisdictional claims in published maps and institutional affiliations.

Article

Wearable Laser Doppler Flowmetry Sensor: A Feasibility Study with Smoker and Non-Smoker Volunteers

Mou Saha [1,*], Viktor Dremin [1,2,*], Ilya Rafailov [3], Andrey Dunaev [2], Sergei Sokolovski [1] and Edik Rafailov [1,4]

[1] Aston Institute of Photonic Technologies, Aston University, Birmingham B4 7ET, UK; s.sokolovsky@aston.ac.uk (S.S.); e.rafailov@aston.ac.uk (E.R.)

[2] Research & Development Center of Biomedical Photonics, Orel State University, 302026 Orel, Russia; dunaev@bmecenter.ru

[3] Aston Medical Technology Ltd., Birmingham B7 4BB, UK; i.e.rafailov@outlook.com

[4] Saratov State University, 410012 Saratov, Russia

[*] Correspondence: m.saha@aston.ac.uk (M.S.); v.dremin1@aston.ac.uk (V.D.)

Received: 29 October 2020; Accepted: 3 December 2020; Published: 7 December 2020

Abstract: Novel, non-invasive wearable laser Doppler flowmetry (LDF) devices measure real-time blood circulation of the left middle fingertip and the topside of the wrist of the left hand. The LDF signals are simultaneously recorded for fingertip and wrist. The amplitude of blood flow signals and wavelet analysis of the signal are used for the analysis of blood perfusion parameters. The aim of this pilot study is to validate the accuracy of blood circulation measurements recorded by one such non-invasive wearable LDF device for healthy young non-smokers and smokers. This study reveals a higher level of blood perfusion in the non-smoker group compared to the smoker group and vice-versa for the variation of pulse frequency. This result can be useful to assess the sensitivity of the wearable LDF sensor in determining the effect of nicotine for smokers as compared to non-smokers and also the blood microcirculation in smokers with different pathologies.

Keywords: wearable laser Doppler flowmetry; blood perfusion; wavelet analysis; smokers

1. Introduction

Tobacco smoking is a major single cause of global cancer deaths and has been labelled as the biggest risk factor for premature deaths in industrialised countries such as the UK [1–3]. It is well known that tobacco smoking directly affects the cardiovascular system [4] through several mechanisms such as atherosclerosis, development of ischemic heart disease, and peripheral artery disease and when combined with other risk factors such as hyperlipidaemia, hypertension and obesity [5]. Nicotine and carbon monoxide, two major constituent chemicals in cigarettes, interfere with the ability of the cardiovascular system to function normally. Exposure to nicotine and carbon monoxide change the heart and blood vessels in ways that increase the risk of heart and cardiovascular disease [5,6]. Nicotine causes human blood vessels to constrict [6], which limits the amount of blood that flows to human organs. It is well known that the blood is the most important connective tissue in the human body that is primarily responsible for the distribution of nutrients and oxygen and the removal of carbon dioxide and metabolic waste products from organs. Any deviation in normal blood circulation, such as over or underflow may indicate a major systemic abnormality. Thus, as an effect of smoking, the constant constriction in the blood vessels results in stiff and non-elastic blood vessels. To compensate for this, the heart starts pumping more blood around the body resulting in its enlargement and increased heart rate [7]. This increased rate, an enlarged heart, and stiffer blood vessels make it harder to pump the blood and provide the body with the required amount of oxygen and nutrients [6]. These structural changes in the blood vessels and heart increase the risk of high blood pressure and cardiovascular disease, ultimately leading to higher mortality and morbidity among the smoking population bringing significant financial burdens on the National Health System and the economy [8,9].

Continuous monitoring of human health and activity using wireless wearable devices will be a key technology in the ubiquitous sensor network society for years to come. Recent advances in the development of the latest-generation of watch-size wearable devices based on ultra-compact semiconductor lasers have opened a new perspective for the implementation of compact blood flow monitoring sensor systems for personalised medicine [10,11]. This wearable device is based on laser Doppler flowmetry (LDF), which is closely related to the dynamic light scattering approach [12]. This technology is widely used for non-invasive measurements in living tissue optical parameters. These devices use a single-mode fibre coupled near-infrared (NIR) laser irradiation which is scattered and reflected from moving red blood cells (RBCs) [10]. Thereafter, these RBC movements generate the frequency-shifted scattering of the initial illumination which is detected by one or two photodetectors with the appropriate signal processing of both photocurrents (initial and reflected). This allows for the subsequent evaluation of the intensity of the blood perfusion. These data are then analysed with the Fourier approach of giving an estimation of the Doppler spectrum that means the LDF records are proportional to the RBCs' velocity. Thus, LDF is used for functional diagnostics of the blood circulation system as well as significant diseases associated with cardiovascular disorders and their complications. Furthermore, this method allows evaluation of the oscillatory processes in the microcirculatory systems. Five rhythmic oscillations are isolated from LDF recordings with the help of wavelet analysis; endothelial (frequency interval 0.0095–0.02 Hz), neurogenic (0.02–0.06 Hz), myogenic (0.06–0.16 Hz), respiratory (0.16–0.4 Hz), and cardiac or pulse rhythm (0.4–1.6 Hz) [13]. Recent studies on these wearable LDF devices have shown the high synchronisation of blood flow rhythms in the contralateral limbs of healthy volunteers [11]. Additionally, it was reported that wearable devices can measure the age-related changes in blood perfusion for healthy participants [10]. Although there were quite a few numbers of articles available based on smoker's blood microcirculation measured by LDF [14–18], no research was conducted in understanding the comparable modulation of the blood flow parameters in non-smoking and smoking volunteers using wearable LDF. Thus, this study demonstrated the feasibility of the wearable device in distinguishing cardiovascular parameters between non-smoking and smoking groups of volunteers. This preliminary research was necessary for planning more expensive and large-scale pre-clinical trials. Within the framework of the work, the strengths and weaknesses of the proposed LDF wearable approach were revealed. The applicability of the sensor for participants was assessed and estimation of the size of the participant groups were evaluated for obtaining sufficient statistical power of future studies.

2. Materials and Methods

2.1. Wearable Laser Doppler Flowmetry Monitor

The experiments were performed with two wearable LDF monitors "FET-1" (Aston Medical Technology Ltd., http://www.amedtech.co.uk/) for recording the blood perfusion (Figure 1). These devices consisting of three identical channels for recording blood perfusion, skin temperature, and movements provide measurement at any desirable point of the human body. The system also comprises a wireless data acquisition module.

Every wearable sensor in the system uses a VCSEL chip (850 nm, 1.4 mW/3.5 mA, Philips, The Netherlands) as a single-mode laser source to implement fibre-free direct illumination of tissue. The fibre probe movements can cause high-frequency intensity fluctuations due to speckle movement. The intensity fluctuations can themselves produce an apparent Doppler shift [19,20] which will highly disturb the initial data acquisition creating faulty conclusions. Fibre-free solution and direct illumination of tissue by the laser diode make it possible to decrease these artefacts which are common in fibre-based LDF systems, as well as to avoid fibre coupling losses. To find a correlation between the changes in the registered blood perfusion and actual body movements the integral accelerometer has been embedded in the sensor.

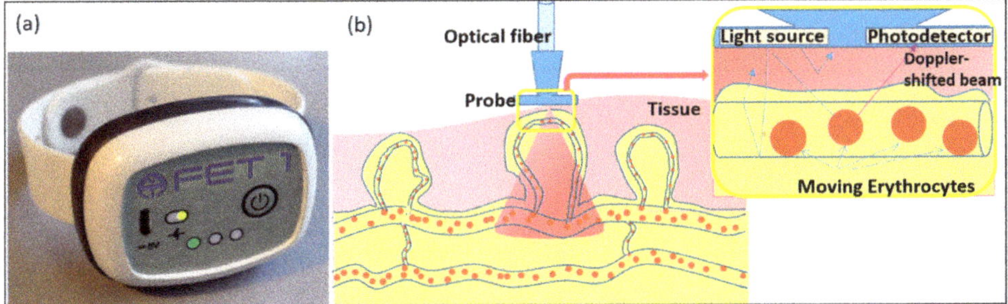

Figure 1. (**a**) Laser Doppler flowmetry (LDF) based wearable device, (**b**) schematic of the principle of LDF technique.

The measurement channel of the devices uses two identical primary signal processing channels. This two-channel scheme provides a more accurate recording of the perfusion value (in particular, the recording of the Doppler shift). Mathematically, the recorded signal is represented as:

$$PU = \int_{\omega_1}^{\omega_2} \omega S(U_1(t) - U_2(t)) d\omega, \tag{1}$$

where PU is perfusion index, $S(U_1(t) - U_2(t))$ is power spectral density of the difference signal from two photodetector channels, $U_1(t), U_2(t)$ is the signal from the photodetector converted to voltage.

The signal is also normalised to the constant component of the photocurrent. The radiation, scattered on moving red blood cells, and the back-reflected signal are received by two photodiodes. Next, the current is converted on the transimpedance amplifier, and the corresponding signal amplification is performed. The next stage of processing is low-pass filtering, constant component extraction, and high-pass filtering. Next, the variable component is normalised to a constant to avoid the influence of different levels of reflection under different optical conditions. The final component is a differential amplifier that protects the circuit from common-mode interference in the signal and reduces the impact of motion artifacts. In general, the presented wearable devices have an identical electronic circuit with stationary LDF devices [13,21,22] and allow the registration of the same signals.

2.2. Study Design

This pilot study was conducted in accordance with the principles set out in the Helsinki Declaration of 2013 by the World Medical Association and was approved by the human ethics committee of Aston University, Birmingham, UK (Ethics approval #1685). This pilot-study involved nine healthy volunteers without pre-existing cardiovascular and other chronic medical conditions who are divided into two groups: non-smokers and smokers. The average age of the non-smoker group was 34.2 ± 3.8 years and the smoker group was 38 ± 4.7 years. Two females and three males were included in the non-smoker group and smoker group consisted of 1 female and 3 males for this pilot study. None of the participants had a history of cardiovascular disease (CVD) or other illness and they were not under any medications. After being informed and explained the study design, every volunteer provided written consent and filled a questionnaire to detail their current health conditions. A short history for the volunteers including medication history, alcohol consumption for the last 24 hours, history of exercise (such as cycling, treadmill, jogging), and a detailed history of smoking were taken.

2.3. Testing Procedure

Blood perfusion parameters were collected in a sitting position, in a state of physical and mental rest (avoid reading, writing, and talking). The hand of the volunteers was placed on a table at the heart's level and they were requested not to ingest any caffeine and alcohol-containing drinks at least one hour and twelve hours prior to allocated measurements time, respectively. The index of blood perfusion was recorded for 8 minutes, while the sensors were attached to the surface of the left middle fingertip and the topside of the left wrist without applying any pressure on the study area. The fingertip was chosen because it is rich in arteriolar venular anastomoses (AVA), whereas the wrist is known to contain less AVA. The measurement was taken twice a day: morning (around 11.00) and evening (around 18.00) for any 5 days in consecutive two weeks.

2.4. Data Proccesing and Analysis

Specialised software was developed to work with the system. This allowed for real-time control of the course of the experiment and analysis of the recorded parameters. Figure 2 depicts an example of the displayed parameters which show the raw data of blood perfusion, temperature, and the movement for the fingertip and wrist in Figure 2a. After acquiring the data, the oscillation rhythms of each measurement were analysed using the built-in module "wavelet analysis" [23] which is displayed in Figure 2b. This wavelet analysis determines the maximum amplitude of blood perfusion and corresponding data for each of the five oscillations mentioned in the previous section.

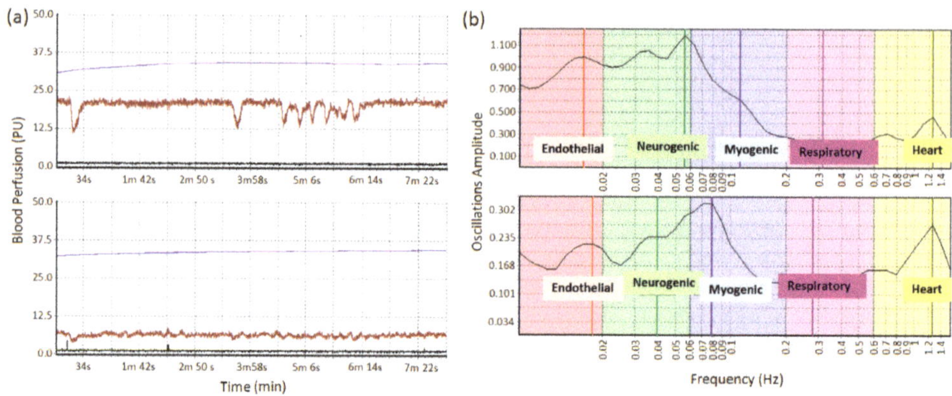

Figure 2. (**a**) Representative recordings of the blood perfusion (in brown), temperature (in blue) and accelerometric movement (in black) for left fingertip (upper) and left wrist (lower), (**b**) representative wavelet analysis for left fingertips (upper) and left wrist (lower).

The Morlet wavelet transform was used for the frequency analysis of registered signals [24–26]. In short, the LDF signal was decomposed using a wavelet transform as:

$$W(s, \tau) = \frac{1}{\sqrt{s}} \int_{-\infty}^{\infty} x(t) \psi^*\left(\frac{t - \tau}{s}\right) dt, \tag{2}$$

where $x(t)$ is a target signal, τ is local time index, s is scaling factor, * means complex conjugation. The Morlet wavelet defined in the form

$$\psi(t) = e^{2\pi i t} e^{-t^2/\sigma} \tag{3}$$

was used with the decay parameter σ = 1. This wavelet allows one to ensure sufficient time-frequency resolution and is well localized in the time domain.

Figure 2 shows that the forearm region has smaller variations of the LDF signal, and spectral analysis shows pronounced peaks in the cardiac, myogenic, neurogenic, and endothelial ranges. We excluded respiratory fluctuations from further statistical analysis due to the lack of pronounced peaks, which may be due to recording signals in basal conditions without any functional tests [27,28].

Taking into account the relatively small sample sizes, nonparametric methods were used to confirm the reliability of differences in the results, namely the Mann–Whitney U-test. Values of $p < 0.01$ were considered significant. Based on the results of this pilot study, we have performed sample size estimations for minimisation of type two error in future studies:

$$n = 2SD^2 \frac{(Z_{\alpha/2} + Z_\beta)^2}{d^2},$$ (4)

where is the SD is standard deviation; $Z_{\alpha/2}$ = 1.96 at type 1 error of 5%; Z_β = 0.84 at 80% power; d is the difference between mean values [29].

3. Results

Figure 3a, b shows the blood perfusion for all the individual participants including non-smokers and smokers at palmer skin of the left middle finger and left wrist, recorded in morning and evening sessions. It does not show any noticeable difference between non-smokers and smokers. After averaging for all non-smokers and smokers, the study reveals a higher level of perfusion in the non-smoker group as compared to the smoker group, showed in Figure 3c,d.

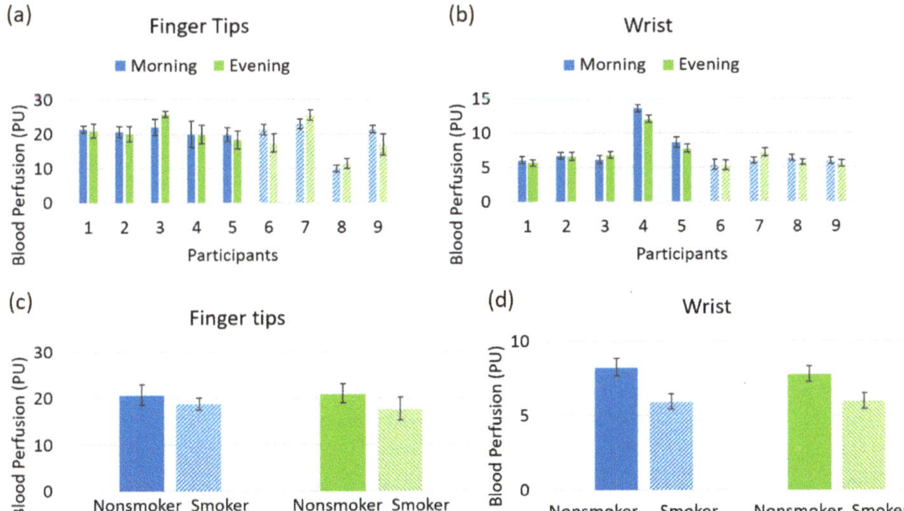

Figure 3. (**a**) Blood perfusion with standard deviation of left fingertip for every participant; (**b**) blood perfusion with standard deviation of left wrist for every participant; (**c**) averaged blood perfusion with standard deviation of fingertip for non-smoker and smoker, (**d**) averaged blood perfusion with standard deviation of wrist for non-smoker and smoker (solid bar denotes non-smoker and patterned bar denotes smoker).

The results depicted in Figures 4 and 5 reveal the comparisons of the wavelet analysis parameters on the fingertip and wrist for both subject groups. The variations of the amplitude of the endothelial (A_e), neurogenic (A_n), myogenic (A_m), and pulse (A_p) are shown in Figure 4a,b for fingertip and wrist respectively. Amplitude measurements on the fingertip are consistently higher for the non-smoker group as compared to the smoker group for all oscillations at both morning and afternoon sessions. However, data received from the wrist does not show any conclusive variations between non-smoker and smoker groups. This is possibly due to the lower number of AVAs in the wrist area resulting in a lack of conclusive information. In addition, pulse amplitude data taken from both fingertip and wrist exhibits a decrease for the smoker group at both morning and afternoon sessions.

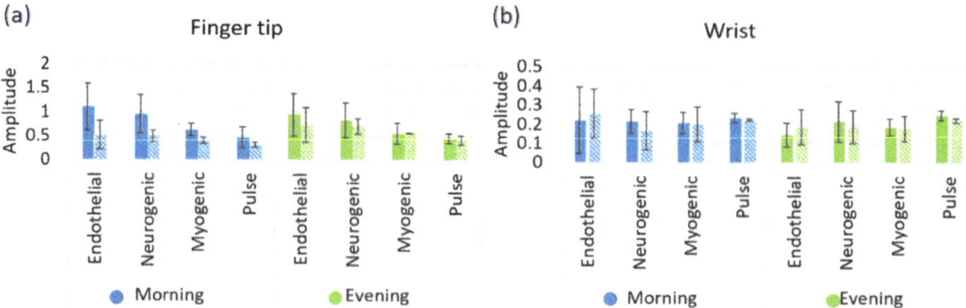

Figure 4. Maximum amplitude with standard deviation of the endothelial, neurogenic, myogenic and pulse mechanism for non-smoker and smoker at (**a**) fingertip, (**b**) wrist. Solid bar denotes non-smoker and patterned bar denotes smoker.

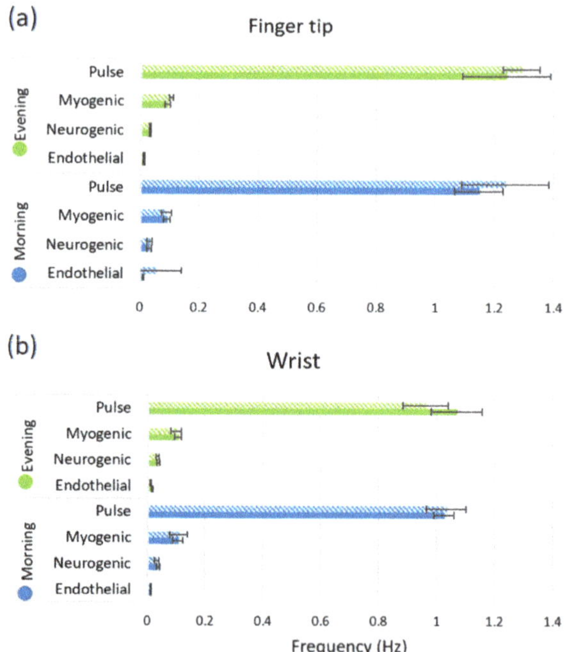

Figure 5. Maximum peak frequency with standard deviation of the endothelial, neurogenic, myogenic and pulse mechanism for non-smoker and smoker at (**a**) fingertip and (**b**) wrist. Solid bar denotes non-smoker and patterned bar denotes smoker.

In Figure 5, the maximum peak frequencies are observed for endothelial, neurogenic, myogenic and pulse rhythms. The oscillation frequency varies for the smokers as compared to non-smokers, shown in Figure 5. It is clearly demonstrated that the pulse frequency is higher for the smoker group as compared to the non-smoker group at both morning and afternoon sessions for fingertip. Additionally, the wrist shows the same pattern for the morning session. This indicates that smoking raises blood pressure and increases the heart rate for smokers, which is confirmed in Figure 5a. For other oscillations, the frequency shows higher value for smoker than non-smoker's fingertip at both morning and afternoon sessions, however, it is slightly different for wrist position.

4. Discussion

Here, we presented a pilot study to assess the feasibility of the wearable device in differentiating cardiovascular parameters between non-smoking and smoking groups of volunteers. Due to the small sample size, statistical analysis did not show significant differences, but the results show clear trends for differences in the measured and calculated parameters.

Lower blood perfusion for the smokers was observed compared to non-smokers in Figure 3c,d. This is probably due to the effect of nicotine, a major constituent of cigarette smoke. Nicotine constricts the blood vessels as well as those in the skin and coronary blood vessels. This vasoconstriction of the skin results in reduced cutaneous blood flow.

The amplitude of endothelial mechanism reduces for the smokers at fingertips, evidenced in Figure 4a. It is known that endothelial cells inside of the blood and heart vessels help to regulate blood clotting and vascular relaxation. These cells generally synthesize and release nitric oxide (NO) which dilate the blood vessels of the body. However, nicotine induces the chance to increase the endothelial dysfunction which is known to cause vasoconstricting substances and narrows the blood vessels [30]. As a consequence, the blood perfusion reduces for the smoker group, evidenced in Figure 3c,d. The plots for fingertip and wrist both show a decrease in amplitude of the neurogenic oscillation for smokers. This is an indicator of higher neurogenic resistance and possible decreased blood flow in the arterioles for the smoker group, which was noticed in the blood perfusion plot of Figure 3c [13]. The source of myogenic oscillation is representative of the spontaneous activity of smooth muscle cells that is associated with the regulation of blood pressure of the human body. Nicotine increases blood pressure for smokers and consequently increases the tension of the vascular wall resulting in the contraction of the vascular smooth muscle. The amplitude of myogenic oscillation decreases for smokers as shown in Figure 4a. Measurements from the wrist show a similar amplitude pattern in myogenic rhythm.

The pulse frequency was higher for the smokers as shown in Figure 5 because the cigarette smoking increases the cardiac work by stimulating the heart rate approximately 10–15 bpm and the blood pressure (acute increase 5–10 mm Hg) [31]. It was already discussed that nicotine constricts the blood vessels, but it can also increase coronary blood flow by increasing the cardiac output, causing subsequent flow-mediated dilation (FMD). It means that the heart needs to pump more due to constriction of the blood vessels that increases the cardiac output as well as heart rate which was evidenced in Figure 5. In addition, the prolonged exposure of carbon monoxide from a cigarette can increase carboxyhemoglobin concentration to as high as 10% for the heavy smoker and may induce anaemia as it binds more readily to haemoglobin than oxygen. As a result, it blocks oxygen-binding sites and impairs the release of the oxygen that is able to bind. Carbon monoxide induced hypoxemia enhances the chance of smoking-related thrombogenesis via increased blood viscosity as the body compensates it by increasing red blood cell mass.

Overall, the current study showed that the wearable device (FET-1) is capable of differentiating cardiovascular parameters between non-smokers and smokers. Our promising results demonstrate the robustness of both the data acquisition and the spectral analysis methods employed to characterise measured optical data. However, it is necessary to continue research in a broader way with a larger sample size to provide a clinical and statistically significant difference in blood perfusion parameters between non-smokers and smokers.

5. Study Limitations and Future Directions

It is shown that most of our results should be considered as preliminary estimates because the research has some limitations. The future work will focus on overcoming the limitations described below.

The number of studied subjects in this research was relatively small. High parameters divergence together with low difference leads to low robustness of the results. Spectral characteristics of LDF samples are characterised by high intra-subject variability. Our estimations have shown that the sample size should be estimated in at least 80 subjects in every group under consideration to minimize type II error.

This work had sufficient limitations due to the LDF sample length, which led to inaccurate interpretation on low frequency oscillations. For reliable statistics, one should ideally include 10 cycles for each of the frequencies under investigation. We had an 8 min recording, which was why the reliable results can be obtained only for frequencies higher than 0.02 Hz. For lower frequencies, the results presented demonstrate only the tendency towards to quantitative data.

6. Conclusions

The present pilot-study demonstrated that the applied wearable device (FET-1) is capable of differentiating cardiovascular parameters between non-smokers and smokers. It presents the results of blood perfusion measurements with the rhythmic oscillations using wearable VCSEL-based sensor system. Studies have shown a comparatively low blood flow for smoker volunteers than the non-smokers. This LDF based wearable sensor system has several advantages which open excellent prospects for a new type of experiments. Power-efficient VCSEL-based devices perform the long-term blood perfusion monitoring which is completely non-invasive. Experiments have shown that the introduction of wireless wearable devices for recording blood microcirculation is a convenient solution for use in medical diagnostics. The wearable implementation of LDF has a high potential in the field of monitoring cardiovascular diseases and is also of great interest for the diagnosis of other conditions associated with microvascular disorders. Portability and low sensitivity to motion artefacts make them mobile and suitable for home use. Furthermore, the sensor device demonstrates the spectral analysis of LDF signal using wavelet transformation to evaluate the regulatory mechanisms and to distinguish the difference between non-smokers and smokers. The presented pilot study of the wearable VCSEL-based LDF sensors provides early results that will need further validation with larger clinical studies.

Author Contributions: Data acquisition and processing, original draft preparation, M.S.; methodology development, data processing, V.D. and A.D.; methodology development, experiments planning and supervision, S.S.; device development, I.R.; conceptualisation and supervision, E.R. All authors have read and agreed to the published version of the manuscript.

Funding: This project has received funding from Engineering and Physical Sciences Research Council (EPSRC) (grant No. EP/R024898/1). Authors also acknowledge support from the Russian Science Foundation (grants No. 18-15-00172) and the funding received within the H2020-MSCA-IF-2018 scheme (grant No. 839888).

Acknowledgments: We would like to thank all the volunteers for their contribution to this pilot study.

Conflicts of Interest: The authors declare no conflict of interest.

References

1. Tobacco Control Local Policy Statement—Cancer Research UK. Available online: https://www.cancerresearchuk. org/sites/default/files/tobacco_control_local_policy_statement.pdf (accessed on 29 October 2020).
2. Branca, F.; Lartey, A.; Oenema, S.; Aguayo, V.; Stordalen, G.A.; Richardson, R.; Arvelo, M.; Afshin, A. Transforming the food system to fight non-communicable diseases. *BMJ* **2019**, *364*, 1296. [CrossRef] [PubMed]
3. Sharma, S. New approaches in smoking cessation. *Indian Heart J.* **2008**, *60*, B34–B37. [PubMed]
4. Blackburn, H.; Brozek, J.; Taylor, H.L. Common circulatory measurements in smokers and nonsmokers. *Circulation* **1960**, *22*, 1112–1124. [CrossRef]

5. Benowitz, N.L.; Burbank, A.D. Cardiovascular toxicity of nicotine: Implications for electronic cigarette use. *Trends Cardiovasc. Med.* **2016**, *26*, 515–523. [CrossRef]

6. Guyton, A.C.; Hall, J.E. *Textbook of Medical Physiology*, 11th ed.; Elsevier Saunders: Philadelphia, PA, USA, 2006; p. 1116.

7. Tayade, M.C.; Kulkarni, N.B. A comparative study of resting heart rate in smokers and nonsmokers. *Int. J. Cur. Res. Rev.* **2012**, *4*, 59–62.

8. Joseph, P.; Leong, D.; McKee, M.; Anand, S.S.; Schwalm, J.-D.; Teo, K.; Mente, A.; Yusuf, S. Reducing the global burden of cardiovascular disease, part 1: The epidemiology and risk factors. *Circ. Res.* **2017**, *121*, 677–694. [CrossRef]

9. Ezzati, M.; Lopez, A.D.; Rodgers, A.; Hoorn, S.V.; Murray, C.J. the Comparative Risk Assessment Collaborating Group. Selected major risk factors and global and regional burden of disease. *Lancet* **2002**, *360*, 1347–1360. [CrossRef]

10. Loktionova, Y.I.; Zherebtsov, E.A.; Zharkikh, E.V.; Kozlov, I.O.; Zherebtsova, A.I.; Sidorov, V.V.; Sokolovski, S.G.; Rafailov, I.E.; Dunaev, A.V.; Rafailov, E.U. Studies of age-related changes in blood perfusion coherence using wearable blood perfusion sensor system. *Proc. SPIE* **2019**, *11075*, 1107507.

11. Zherebtsov, E.A.; Zharkikh, E.V.; Kozlov, I.O.; Zherebtsova, A.I.; Loktionova, Y.I.; Chichkov, N.B.; Rafailov, I.E.; Sidorov, V.V.; Sokolovski, S.G.; Dunaev, A.V.; et al. Novel wearable VCSEL-based sensors for multipoint measurements of blood perfusion. *Proc. SPIE* **2019**, *10877*, 1087708.

12. Fredriksson, I.; Fors, C.; Johansson, J. *Laser Doppler Flowmetry-a Theoretical Framework Department of Biomedical Engineering*; Linköping University: Linköping, Sweden, 2007; pp. 1–22.

13. Dunaev, A.V.; Sidorov, V.V.; Krupatkin, A.I.; Rafailov, I.E.; Palmer, S.G.; Stewart, N.A.; Sokolovski, S.G.; Rafailov, E.U. Investigating tissue respiration and skin microhaemocirculation under adaptive changes and the synchronization of blood flow and oxygen saturation rhythms. *Physiol. Meas.* **2014**, *35*, 607. [CrossRef]

14. Fushimi, H.; Kubo, M.; Inoue, T.; Yamada, Y.; Matsuyama, Y.; Kameyama, M. Peripheral vascular reactions to smoking—Profound vasoconstriction by atherosclerosis. *Diabetes Res. Clin. Pract.* **1998**, *42*, 29–34. [CrossRef]

15. Uehara, K.; Sone, R.; Yamazaki, F. Cigarette smoking following a prolonged mental task exaggerates vasoconstriction in glabrous skin in habitual smokers. *J. UOEH* **2010**, *32*, 303–316. [CrossRef] [PubMed]

16. Meekin, T.; Wilson, R.; Scott, D.; Ide, M.; Palmer, R. Laser Doppler flowmeter measurement of relative gingival and forehead skin blood flow in light and heavy smokers during and after smoking. *J. Clin. Periodontol.* **2000**, *27*, 236–242. [CrossRef] [PubMed]

17. Palmer, R.; Scott, D.; Meekin, T.; Poston, R.; Odell, E.; Wilson, R. Potential mechanisms of susceptibility to periodontitis in tobacco smokers. *J. Periodontal Res.* **1999**, *34*, 363–369. [CrossRef] [PubMed]

18. Tur, E.; Yosipovitch, G.; Oren-Vulfs, S. Chronic and acute effects of cigarette smoking on skin blood flow. *Angiology* **1992**, *43*, 328–335. [CrossRef]

19. Newson, T.P.; Obeid, A.; Wolton, R.S.; Boggett, D.; Rolfe, P. Laser Doppler velocimetry: The problem of fibre movement artefact. *J. Biomed. Eng.* **1987**, *9*, 169–172. [CrossRef]

20. Rajan, V.; Varghese, B.; van Leeuwen, T.G.; Steenbergen, W. Review of methodological developments in laser Doppler flowmetry. *Lasers Med. Sci.* **2009**, *24*, 269–283. [CrossRef]

21. Dremin, V.; Zherebtsov, E.; Sidorov, V.; Krupatkin, A.; Makovik, I.; Zherebtsova, A.; Zharkikh, E.; Potapova, E.; Dunaev, A.; Doronin, A.; et al. Multimodal optical measurement for study of lower limb tissue viability in patients with diabetes mellitus. *J. Biomed. Opt.* **2017**, *22*, 085003. [CrossRef]

22. Makovik, I.N.; Dunaev, A.V.; Dremin, V.V.; Krupatkin, A.I.; Sidorov, V.V.; Khakhicheva, L.S.; Muradyan, V.F.; Pilipenko, O.V.; Rafailov, I.E.; Litvinova, K.S. Detection of angiospastic disorders in the microcirculatory bed using laser diagnostics technologies. *J. Innov. Opt. Heal. Sci.* **2017**, *11*, 1750016. [CrossRef]

23. Goltsov, A.; Anisimova, A.V.; Zakharkina, M.; Krupatkin, A.I.; Sidorov, V.V.; Sokolovski, S.G.; Rafailov, E. Bifurcation in blood oscillatory rhythms for patients with ischemic stroke: A small scale clinical trial using laser doppler flowmetry and computational modeling of vasomotion. *Front. Physiol.* **2017**, *8*, 160. [CrossRef]

24. Tankanag, A.; Chemeris, N. Application of the adaptive wavelet transform for analysis of blood flow oscillations in the human skin. *Phys. Med. Biol.* **2008**, *53*, 5967. [CrossRef] [PubMed]

25. Stefanovska, A.; Bracic, M.; Kvernmo, H.D. Wavelet analysis of oscillations in the peripheral blood circulation measured by laser Doppler technique. *IEEE Trans. Biomed. Eng.* **1999**, *46*, 1230–1239. [CrossRef] [PubMed]

Biosensors **2020**, *10*, 201

26. Dremin, V.; Kozlov, I.; Volkov, M.; Margaryants, N.; Potemkin, A.; Zherebtsov, E.; Dunaev, A.; Gurov, I. Dynamic evaluation of blood flow microcirculation by combined use of the laser Doppler flowmetry and high-speed videocapillaroscopy methods. *J. Biophotonics* **2019**, *12*, e201800317. [CrossRef] [PubMed]

27. Krasnikov, G.V.; Tyurina, M.Y.; Tankanag, A.V.; Piskunova, G.M.; Chemeris, N.K. Analysis of heart rate variability and skin blood flow oscillations under deep controlled breathing. *Respir. Physiol. Neurobiol.* **2013**, *185*, 562–570. [CrossRef] [PubMed]

28. Trembach, N.; Zabolotskikh, I. Breath-holding test in evaluation of peripheral chemoreflex sensitivity in healthy subjects. *Respir. Physiol. Neurobiol.* **2017**, *235*, 79–82. [CrossRef]

29. Charan, J.; Biswas, T. How to calculate sample size for different study designs in medical research? *Indian J. Psychol. Med.* **2013**, *35*, 121–126. [CrossRef]

30. Endothelial Dysfunction. Available online: https://www.diabetes.co.uk/diabetes-complications/endothelial-dysfunction.html (accessed on 29 October 2020).

31. U.S. Department of Health and Human Services. *How Tobacco Smoke Causes Disease: The Biology and Behavioral Basis for Smoking-Attributable Disease: A Report of the Surgeon General.* Available online: https://www.ncbi.nlm.nih.gov/books/NBK53017/ (accessed on 29 October 2020).

Publisher's Note: MDPI stays neutral with regard to jurisdictional claims in published maps and institutional affiliations.

Article

Demonstration of a Label-Free and Low-Cost Optical Cavity-Based Biosensor Using Streptavidin and C-Reactive Protein

Donggee Rho and Seunghyun Kim *

Electrical and Computer Engineering Department, Baylor University, One Bear Place #97356, Waco, TX 76798, USA; donggee_rho@baylor.edu
* Correspondence: Seunghyun_Kim@baylor.edu

Abstract: An optical cavity-based biosensor (OCB) has been developed for point-of-care (POC) applications. This label-free biosensor employs low-cost components and simple fabrication processes to lower the overall cost while achieving high sensitivity using a differential detection method. To experimentally demonstrate its limit of detection (LOD), we conducted biosensing experiments with streptavidin and C-reactive protein (CRP). The optical cavity structure was optimized further for better sensitivity and easier fluid control. We utilized the polymer swelling property to fine-tune the optical cavity width, which significantly improved the success rate to produce measurable samples. Four different concentrations of streptavidin were tested in triplicate, and the LOD of the OCB was determined to be 1.35 nM. The OCB also successfully detected three different concentrations of human CRP using biotinylated CRP antibody. The LOD for CRP detection was 377 pM. All measurements were done using a small sample volume of 15 µL within 30 min. By reducing the sensing area, improving the functionalization and passivation processes, and increasing the sample volume, the LOD of the OCB are estimated to be reduced further to the femto-molar range. Overall, the demonstrated capability of the OCB in the present work shows great potential to be used as a promising POC biosensor.

Keywords: biosensors; optical cavity-based biosensor; biomarker detection

Citation: Rho, D.; Kim, S. Demonstration of a Label-Free and Low-Cost Optical Cavity-Based Biosensor Using Streptavidin and C-Reactive Protein. *Biosensors* **2021**, *11*, 4. https://dx.doi.org/10.3390/bios11010004

Received: 7 December 2020
Accepted: 22 December 2020
Published: 24 December 2020

Publisher's Note: MDPI stays neutral with regard to jurisdictional claims in published maps and institutional affiliations.

1. Introduction

The early diagnosis of diseases, including cancers, infectious diseases, and cardiovascular diseases, is vital in order to apply effective treatments and increase the chance of full recovery [1–4]. Diagnostic technologies in the current healthcare system are mostly used at centralized laboratories, involve costly and time-consuming processes, and are operated by expert staff [2,5,6]. For example, enzyme-linked immunosorbent assay (ELISA), considered as the gold standard diagnostic method, is labor-intensive, requiring complicated procedures such as labeling and multiple washing steps [2,5,7,8]. Label-free optical biosensing methods such as surface plasmon resonance (SPR) and total internal reflection ellipsometry (TIRE) have been extensively investigated and developed [9,10]. SPR and TIRE biosensors are label-free biosensors without complex procedures, and are highly sensitive with reduced assay times. However, some drawbacks, including high-cost, bulky size, and the need for trained personnel to operate, remain to be improved [11,12]. With these limitations in the current diagnostic technologies, it is difficult for people to monitor their health status regularly, which would eventually increase the chance of being diagnosed with diseases at late stages [3,7]. The problems become worse for people who are in financial difficulties and living in developing countries with deficient healthcare systems [6,13]. To address these challenges existing in the conventional diagnostic field, a point-of-care (POC) biosensor has emerged as a promising alternative, allowing patients to regularly check their health condition at the bedside or near them without being dependent on

Biosensors **2021**, *11*, 4. https://dx.doi.org/10.3390/bios11010004   https://www.mdpi.com/journal/biosensors

central laboratory testing [4,6,14–17]. According to the World Health Organization, an ideal POC test should satisfy the ASSURED (affordable, sensitive, specific, user-friendly, rapid, equipment-free, deliverable to end-users) criteria [16,17]. One of the most widely available and commercialized POC devices is based on lateral flow assays (LFAs) with their low cost, ease of use, and speed [4,6,18]. However, limitations still remain with regard to LFAs in terms of producing reproducible and sensitive test results [6,18,19].

An optical cavity-based biosensor (OCB) using a differential detection method has been developed for the application of POC diagnostics [20–27]. The structure of an optical cavity consists of two partially reflective surfaces in parallel, separated by a small gap in between. The light propagating through the optical cavity experiences multiple beam interference due to those two reflective surfaces, and produces a transmission spectrum with a resonant characteristic. Because of the resonant characteristic, the optical cavity can be used to detect small changes inside the cavity which, in turn, makes it an attractive platform for biosensing applications. To use the optical cavity for biosensing, the sensing area is functionalized with receptor molecules. When target biomolecules are adsorbed into the receptors, a shift in the resonant response occurs. To detect the small resonant response shift, the OCB measures the changes in optical intensities at two different wavelengths using low-cost laser diodes and a CMOS camera instead of using an expensive spectrometer or a tunable laser, lowering the total cost. The sensitivity of the OCB is enhanced by employing a differential detection method. We designed the optical cavity structure so that the optical intensities of two wavelengths are changing in opposite directions upon the capture of target biomolecules, in order to have a significantly bigger change from the calculated differential values. The differential detection method not only increases the sensitivity but also offers other benefits for biosensing, such as power equalization (no initial power variation depending on the measurement results), a larger dynamic range (the detectable concentration range of the biomolecules), and a larger fabrication tolerance [23,28,29]. The intensity measurement method also enables the simultaneous detection of multiple analytes by immobilizing corresponding bioreceptors at different areas of the optical cavity surface where the laser beams pass through. The capability of this OCB to detect small changes in bulk refractive index was demonstrated by using refractive index fluids with proven portability [25,27]. As a preliminary test to confirm the application of OCB in detecting the binding events at the optical cavity sensing area, the attachment of biotinylated bovine serum albumin (BSA) was measured on the streptavidin-functionalized surface [26].

In this present work, we demonstrate the OCB with streptavidin and C-reactive protein (CRP), and determine the limit of detection (LOD) for these. The optimized optical cavity design with simulations, surface functionalization steps, testing procedures, and measurement results are discussed. We report the use of the OCB for biomarker (CRP) detection for the first time.

2. Materials and Methods

2.1. Materials

(3-Aminopropyl) triethoxysilane (99%, APTES), streptavidin (lyophilized solid), and bovine serum albumin (lyophilized powder, BSA) were purchased from Sigma Aldrich, Inc, St. Louis, USA. Sulfo-NHS-Biotin (EZ-Link, powder) was purchased from Thermo Scientific, Inc, Waltham, USA. Tris-HCl buffer (1M, pH 8.0) was purchased from Bio Basic, Inc, Amherst, USA. Biotin-conjugated C-reactive protein antibody was purchased from Novus Biologicals, LLC, Centennial, USA. Human C-reactive protein (≥97%, CRP) was purchased from R&D Systems, Inc, Minneapolis, USA. Spin-on-glass (IC1-200, SOG) was purchased from Futurrex, Inc, Franklin, USA. SU8 photoresist (SU8-2005) was purchased from Kayaku Advanced Materials, Inc, Westborough, USA. UV glue (NOA 86H) was purchased from Norland Products, Inc, East Windsor, USA.

2.2. Schematic

A schematic of the OCB is shown in Figure 1a. Two low-cost laser diodes at different wavelengths are used as light sources operated by laser diode drivers with the constant current mode. The laser beams are collimated, combined by a 50:50 beam splitter (BS), and alternatively propagate with one-second intervals using a rotating beam blocker. A neutral filter (NF) is placed in the light path to attenuate the intensities of laser beams in order to avoid the saturation of a CMOS camera (Discovery M15, Tucsen). The intensities of laser beams, transmitted through an optical cavity sample (OCS), are measured by the CMOS camera in real-time. Figure 1b shows each layer of the OCS structure, while Figure 1c shows the cross section of it. The bottom and patterned top silver layers on 3-inch glass substrates act as partially reflective surfaces. Spin-on-glass (SOG) layers are spin-coated on top of the silver layers to protect them from possible damages during the sample fabrication process and test, to facilitate the silanization-based surface functionalization process, and to maximize the sensitivity. The microfluidic channel between SOG layers is created by SU8 patterns. The receptor molecules are functionalized at the center area of the microfluidic channel, creating a sensing area. UV glue is used to bond the two separately processed substrates at the end of the fabrication process. To introduce fluids to the OCS without air bubbles, a syringe pump is used to add drops of fluids into the 3D printed input port (volume capacity: 20 µL) through a bent syringe tip, while a low-cost vacuum pump is attached to the 3D printed output port through tygon tubing to pull fluids from the input port through the microfluidic channel.

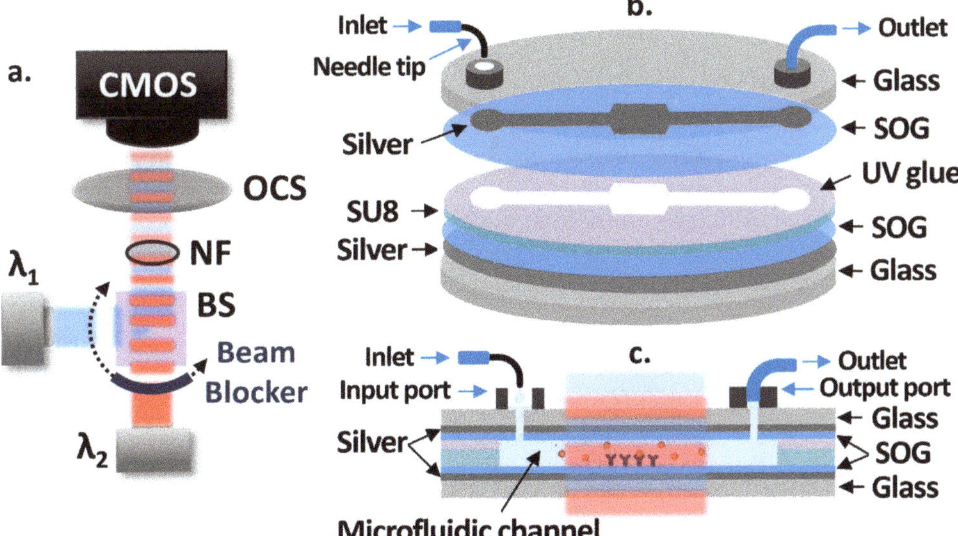

Figure 1. (**a**) Schematic diagram of the optical cavity-based biosensor (OCB) showing two laser beams at two different wavelengths (λ_1 (**blue**) and λ_2 (**red**)) alternatively propagating through the OCS with an interval of one second and reaching the CMOS camera. (**b**) Structure of the OCS showing each layer and connected input and output ports. (**c**) Cross-sectional view of the OCS showing the target biomolecule in a sample fluid being introduced into the microfluidic channel and attached to the receptor molecules on the SOG surface.

2.3. Simulation Results

As illustrated in Figure 1c, the target biomolecules in the sample fluid attach to the receptor molecules which, in turn, causes output intensity changes in the two laser diodes. For simulations, we employed the fixed index model enabling the approximation of the number of the biomolecules attached to receptors on the sensing area to a sensing layer thickness with a fixed refractive index [30]. We set the refractive index of the sensing layer to 1.45, which has been widely accepted for various biomolecules such as proteins, DNAs, and viruses [30–33]. FIMMWAVE/FIMMPROP (Photon design) was used to perform the simulations. For benefits such as enhanced sensitivity, power equalization, a larger dynamic range, and a larger fabrication tolerance, we employed a differential detection method [23,25]. For the differential detection method, a differential value (η) is calculated by the equation below.

$$\eta = \frac{I_1 - I_{10}}{I_{10}} - \frac{I_2 - I_{20}}{I_{20}}$$

I_1 and I_2 are the intensities (efficiencies for simulations) of λ_1 and λ_2, respectively, and I_{10} and I_{20} are the initial intensities for I_1 and I_2, respectively [26].

To achieve the largest differential value change, we searched for the optimal cavity width at which the efficiencies of two different wavelengths (out of available low-cost laser diodes in the market) change the most in the opposite directions with the sensing layer thickness increase. Since many different possible solutions exist, we narrowed our search to use a silver thickness of 20 nm with the microfluidic channel height ranging between 5 μm and 10 μm. We chose this channel height range to limit the fluid volume required to fill the channel without significant fluid flow resistivity. Since the fluid flow resistivity is inversely proportional to the third power of the height, if the height is too small, then the flow rate is slower, and a stronger vacuum pump is necessary to handle the fluids. Considering that the typical size of proteins is less than 20 nm, we focused on the simulation for a sensing layer thickness up to 20 nm. From the simulation results, we found that the differential value change depends on the SOG thickness. This means the local refractive index change due to the sensing layer change is more influential on the resonant characteristic at certain locations inside the cavity, which must be related to the electromagnetic field distribution in the cavity. Based on the spin curve of the SOG, we considered the SOG thickness in the range of 150 nm to 450 nm.

The simulation results for the optimized optical cavity structure are shown in Figure 2. For the wavelengths of 830 nm (λ_1) and 904 nm (λ_2), the optimized optical cavity design has a cavity width (silver-to-silver distance) of 8 μm, and an SOG thickness of 400 nm with a silver thickness of 20 nm. As the sensing layer increased from 0 to 20 nm, the efficiency of 830 nm decreased from 0.18 to 0.137 (−0.043), while the efficiency of 904 nm increased from 0.063 to 0.077 (+0.014). With this opposite changing trend of two wavelengths, the corresponding differential value changed from 0 to 0.481, showing a significantly larger change compared to the individual wavelengths, with a better linearity.

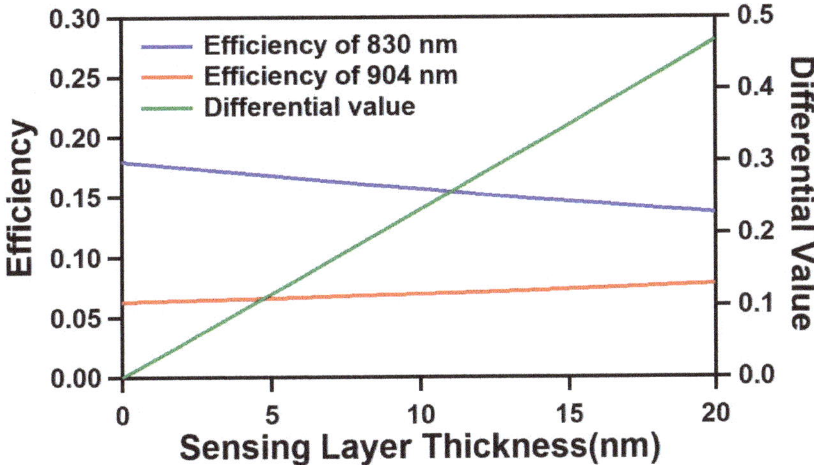

Figure 2. Simulation results showing efficiencies of 830 nm (**blue**) and 904 nm (**red**) wavelengths and differential values (**green**) versus the sensing layer thickness in the range between 0 and 20 nm.

2.4. Sample Fabrication and Surface Functionalization Processes

The fabrication process of the OCS is straightforward without complex micro and nano fabrication steps. First, a 3-inch glass substrate was drilled using a 1 mm diamond drill bit to make the inlet and outlet of a microfluidic channel. A silver layer was sputter-deposited on the drilled glass substrate and on another plain glass substrate. The top silver layer was patterned through a photolithography process followed by a wet-etch process, as shown in Figure 1b, for allowing UV illumination on the UV epoxy to cure and bond two substrates. Then, SOG was spin-coated at 1200 RPM on the silver layer of both substrates and cured at 130 °C on a hot plate. On top of the SOG layer of the plain glass substrate, an SU8 layer was patterned using a photolithography process to define the microfluidic channel. Finally, we used a UV curable epoxy to bond the drilled and plain glass substrates in order to form an optical cavity structure [26]. A top-view image of the fabricated optical cavity microfluidic channel is shown in Figure 3a. The typical layer thicknesses of fabricated devices are, on average, 22 nm (silver), 410 nm (SOG), 6.4 µm (SU8), and 1.08 µm (UV glue). The microfluidic channel has a total length of 5 cm, a height of 7.5 µm (distance between SOG surfaces), and a width of 500 µm, while the width at the sensing area is 1 mm. The sensing area is 2.5 mm^2, and the dotted circular area at the center with a diameter of 160 µm represents the area used for the data processing, calculating the average intensities and differential values.

Figure 3b illustrates the functionalization steps on the SOG surface on the drilled substrate. The oxygen plasma treatment was applied for 5 min to create hydroxyl groups on the SOG surface. We performed the vapor-phase deposition of APTES by placing a substrate in a desiccator with 0.5 mL of 99% APTES in a small container placed at the bottom [34]. The entire desiccator was placed on a hot plate at 90 °C for 24 h to create terminal amine groups (-NH$_2$) on the surface. After the overnight incubation, unbound APTES molecules were removed with DI water in an ultrasonic bath for 7 min, and the glass substrate was baked at 110 °C for 10 min. To functionalize the sensing area, 5 mg/mL of sulfo-NHS-biotin mixed in DI water was applied using a micropipette. It was then incubated for 1 h to covalently immobilize the biotin on the surface through amide bonds, while other areas were passivated with 1% BSA. The surface was then ready for the streptavidin detection experiment. BSA was also applied to the plain substrate with the SU8 pattern to passivate the bottom and side walls of the channel, so as to minimize the nonspecific binding of streptavidin and other biomolecules.

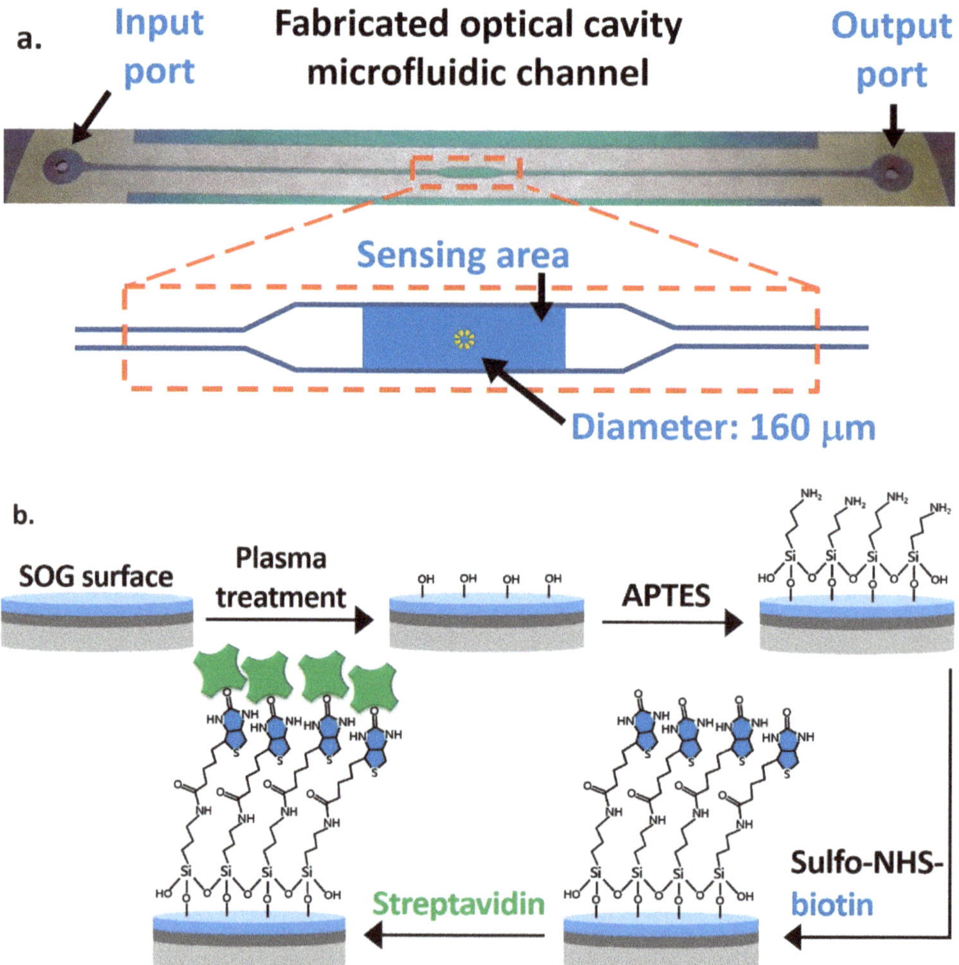

Figure 3. (a) Top view of the fabricated optical cavity microfluidic channel indicating the sensing area at the center and the area for the data process. (b) The functionalization procedure of the spin-on-glass (SOG) surface for the immobilization of streptavidin.

2.5. Test Setup

Figure 4 shows the test setup on an optics table for experiments. Two laser diodes at 830 nm and 904 nm wavelengths were attached to collimators and mounted with kinematic mounts. A 3D-printed beam blocker with a servo motor was located on top of a 50:50 beam splitter to block the laser diode beams alternately. The side view in Figure 4 shows a BS, an NF, a 45-degree mirror under the fabricated OCS in a 3D-printed sample holder, and a CMOS camera. The total cost to build the whole system was low, at about USD 1100 excluding posts, mounts, and the syringe pump.

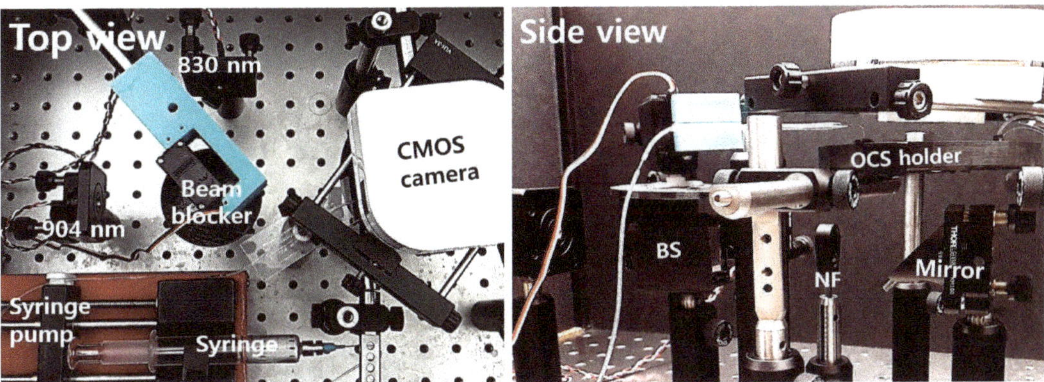

Figure 4. Top and side views of the test setup for the optical cavity-based biosensor (OCB).

2.6. Fine-Tuning of the Optical Cavity Width Using Polymer Swelling

The optical response of any type of optical resonator is very sensitive to its cavity or resonator size. Due to possible errors during the fabrication process, the cavity widths of the fabricated OCSs show some variations. Even with a larger fabrication tolerance using the differential detection method, it is challenging to successfully fabricate the OCSs with a width accurate to within 40 nm, which leads to a low success rate in producing measurable samples [25]. We overcame this problem using the polymer swelling property [35–37]. The microfluidic channel walls were formed of SU8 and UV epoxy polymers. When the microfluidic channel was filled with DI water, the SU8 and UV epoxy in contact with DI water slowly swelled over time. As the optical cavity size increased due to the swelling, the optical intensities changed over time, following the resonance curve. At the optimal cavity width, it was anticipated that the intensities of the two wavelengths would change in opposite directions. During the swelling period, we monitored the optical intensity changes of both wavelengths and conducted sensing experiments when the optical cavity reached this optimal condition. The time it takes for this fine-tuning process varies from less than 1 h to more than 10 h, as the initial optical cavity widths differ. The swelling rate is rapid at the beginning and then slows over time. With this fine-tuning process, we were able to achieve a very high success rate of producing measurable samples (>90%).

3. Results

3.1. Streptavidin Detection

Streptavidin is a 52.8 kDa protein with a dimension of 5.6 nm × 4.2 nm × 4.2 nm [38]. For a monolayer of streptavidin with a height of 5.6 nm, the simulated differential value change due to this monolayer is 0.1266. For measurements, DI water was introduced first through the microfluidic channel for optical cavity width fine-tuning. When the optical cavity reached the measurable condition, 15 μL of streptavidin was then introduced with a flow rate of about 0.9 μL/min for about 17 min. Finally, the channel was rinsed with 15 μL of DI water. Representative trends of four different concentrations of streptavidin in DI water, 300 ng/mL, 1 μg/mL, 3 μg/mL, and 10 μg/mL, are shown in Figure 5 along with the negative control. The average differential value for 2 min before the introduction of streptavidin was set to 0 as the baseline. The change in the differential value due to the binding of streptavidin was measured by averaging differential values between 25 and 27 min, which is 8–10 min after the DI water was introduced for rinsing. For the negative control (black), the sensing area of this channel was blocked with BSA everywhere without sulfo-NHS-biotin. As expected, when 1 μg/mL of streptavidin was introduced into the negative control channel, no obvious change in the differential value was found, while there were some fluctuations during the period with streptavidin in between 0 and 17 min.

This could have been due to the non-specific interaction of streptavidin with BSA. Clearly, some loosely attached streptavidin molecules were removed in the DI water rinse, and the differential values stabilized. The differential value change for the negative control was −0.00213, and the standard deviation was 0.00155. For the 10 µg/mL concentration (yellow), the differential value started changing in 2 min and reached 0.074 in 5.5 min with a slope of 0.0235/min after the introduction of streptavidin (at $t = 0$). The change slowed down from 5.5 min but kept increasing up to 0.095 with a slope of 0.00163/min until the DI water was introduced (at $t = 17$ min) for the rinse. As soon as the channel was rinsed, the change decreased slightly and reached 0.085 (at $t = 27$ min) on average. The result for the concentration of 3 µg/mL (green) shows that the differential value started slowly increasing at around 5 min and reached to 0.04 with a slope of 0.0023/min. The change stopped for about 3 min after the DI water rinse and then increased again to 0.055 with a slower slope of 0.0014/min. This could have been due to the binding of residual streptavidin molecules on the surface during the DI water rinse. The changes in differential values for the streptavidin concentrations of 1 µg/mL and 300 ng/mL started at 8 min and 12 min, respectively, with slower slopes (1 µg/mL: 0.00176/min; 300 ng/mL: 0.00089/min). After the introduction of DI water, they showed a similar trend whereby the differential value decreased for about 3 min and increased for the rest of the 7 min, reaching to 0.027 and 0.16, respectively. Those changes after the DI water rinse could also have been caused by the unbound molecules.

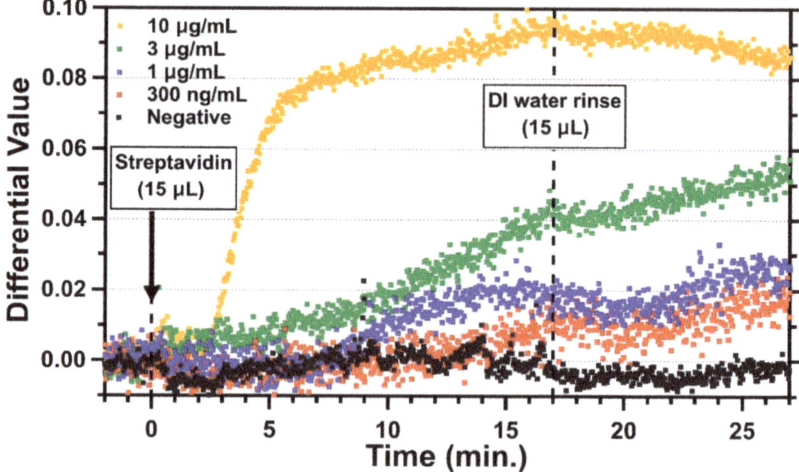

Figure 5. Real-time measurements for 30 min showing the changing differential values after the introduction of 15 µL of streptavidin for four different concentrations and the negative control (10 µg/mL: yellow; 3 µg/mL: green; 1 µg/mL: blue; 300 ng/mL: red; and negative control: black).

The triplicate test results of four different concentrations are shown in Figure 6. The differential value due to the binding of streptavidin was measured by averaging differential values between 25 and 27 min, as described earlier. The average standard deviation of DI water was measured to be 0.00274. The average differential value changes were 0.074 ± 0.018 (10 µg/mL), 0.039 ± 0.0091 (3 µg/mL), 0.024 ± 0.003 (1 µg/mL), and 0.013 ± 0.001 (300 ng/mL). The LOD of our OCB biosensor was determined by the average sensor response crossing the 3σ line (0.00821), which was 71.3 ng/mL (1.35 nM). The differential value of 10 µg/mL did not reach the anticipated value for a monolayer of streptavidin (0.1266). There are a few possible hypotheses to explain this: (1) the streptavidin molecules on the sensing area were oriented towards where the smaller side of the molecule was in the beam propagation direction; (2) the actual refractive index change due to the immobilization of the streptavidin was less than the monolayer with the refractive index of 1.45 used for simulation; or (3) the functionalization and passivation processes were not sufficient to allow streptavidin molecules to form a densely-packed monolayer on the sensing area without losing them through non-specific bindings on other passivated surfaces. Out of these, the third is most likely. Even if the layer created by the immobilized streptavidin molecules was thinner with a lower refractive index, and the differential value for a monolayer of streptavidin was about 0.074 (average differential value change for 10 µg/mL), it is clear we were not able to form a densely packed streptavidin only on the sensing area. Based on the given size of the streptavidin, the total amount of streptavidin required to form a monolayer covering the entire sensing area of 2.5 mm^2 is estimated to be 12.4 ng. For the streptavidin concentration of 1 µg/mL, the total amount of streptavidin in 15 µL of sample fluid is 15 ng. This indicates that, if all available streptavidin molecules are attached densely only on the sensing area, then there are more molecules than are necessary to form a monolayer. If a monolayer is formed and the assumption of the differential value change (0.074) for a monolayer of streptavidin is valid, then the differential value is supposed to reach that level with 1 µg/mL. However, since the differential value change for 1 µg/mL of concentration was only 0.024, the result clearly shows no monolayer was formed. This suggests the sensing area was not well functionalized with active biotin, and/or we lost many streptavidin molecules in other areas. If we improve the functionalization and passivation processes to block other areas of target biomolecules from being attached, so that all available target molecules are attached densely only on the sensing area, the result can be significantly improved.

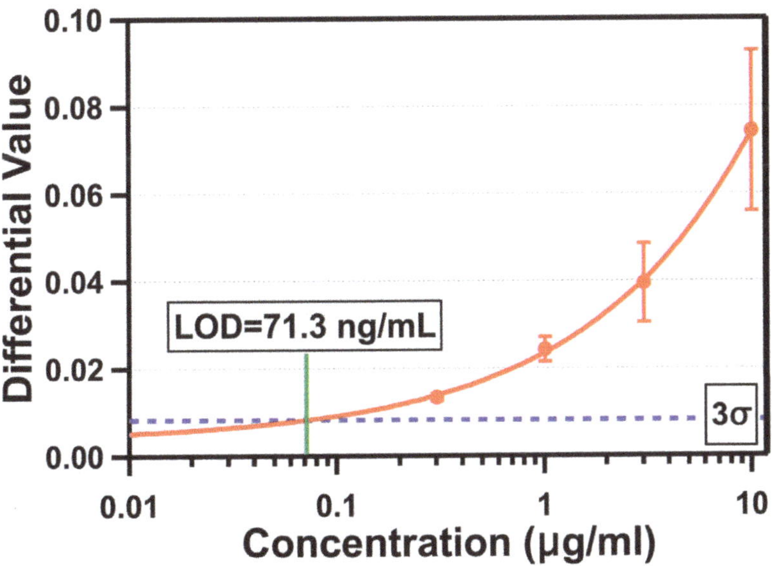

Figure 6. Differential values measured in triplicate versus four concentrations of streptavidin in a log scale.

3.2. C-Reactive Protein (CRP) Detection

CRP is a 115 kDa serum protein with a hydrated volume of 197.3 mm^3, and is one of the most frequently used cardiac biomarkers with high specificity to diagnose and monitor cardiovascular diseases (CVDs), which are the leading cause of death worldwide [39]. The American Heart Association (AHA) and the Center for Diseases Control and Prevention (CDC) defined the risks of CVDs to be low for a concentration of CRP in humans below 1 µg/mL, moderate for a CRP concentration between 1 and 3 µg/mL, and high for a CRP concentration over 3 µg/mL [40]. The level of human CRP is also increased 1000-fold within 24–48 h in response to infection, inflammation, and tissue damage [41]. Figure 7 shows the measurement results for three different concentrations of CRP (10 µg/mL, 1 µg/mL, and 100 ng/mL) using the OCB. For CRP detection, we followed the same fabrication and functionalization processes used for the streptavidin detection. To functionalize the sensing area with the CRP antibody, we first introduced 30 µL of streptavidin with a concentration of 100 µg/mL to the microfluidic channel and incubated for at least 30 min so that the streptavidin molecules were immobilized on the biotin on the sensing area. After rinsing the channel with DI water to remove unbound streptavidin molecules, 30 µL of biotin-conjugated CRP antibody with a concentration of 100 µg/mL was introduced and incubated for at least 30 min so that the biotin part of it was attached to the streptavidin-coated surface, while the CRP antibody covered the surface. The microfluidic channel was rinsed with DI water to remove unbound CRP antibody molecules and filled with it for fine-tuning the optical cavity through polymer swelling. When the optical cavity was ready for the experiments, we introduced 15 µL of CRP protein spiked in tris-HCl buffer with a flow rate of about 0.9 µL/min for about 17 min. Finally, the microfluidic channel was rinsed with DI water, and the average differential value changes were determined by averaging differential values between 8 and 10 min after DI water was introduced. The measured average differential value changes were 0.141 (10 µg/mL), 0.061 (1 µg/mL), and 0.018 (100 ng/mL). Based on the measured average standard deviation in the baseline data with DI water ($\sigma = 0.00284$), the LOD for CRP detection is determined to be 43.3 ng/mL (377 pM).

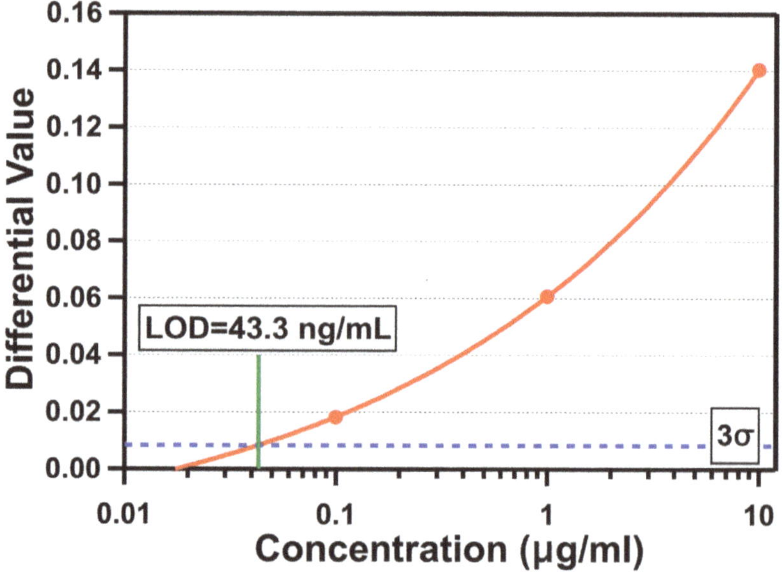

Figure 7. Differential values versus three concentrations of C-reactive protein (CRP) in a log scale.

4. Discussion and Future Work

The LOD of our OCB for streptavidin detection (1.35 nM) can be improved further. First, the sensing area (2.5 mm^2) where streptavidin was allowed to be attached was larger than the area that was used for data processing (0.02 mm^2; the area of a 160 µm diameter circle). If we properly functionalize only this area with sulfo-NHS-biotin, so that streptavidin molecules can be attached within the area of 0.02 mm^2, then the LOD of the OCB becomes 10.9 pM, assuming streptavidin molecules are attached on the sensing area uniformly. Second, as we discussed earlier, the current functionalization and passivation processes need to be improved further to allow more streptavidin molecules to be attached only at the sensing area. If we consider the sensing area of 0.02 mm^2 and lose only about

20% of the target molecules (equivalently, 15 μL of 1 μg/mL streptavidin forms a monolayer on 2.5 mm^2), then we can estimate that the LOD can be improved further by an order of magnitude, to 1.09 pM. Finally, the LOD will be improved proportionally by the amount of sample fluid. This is simply because, to be able to detect the differential value changes, we need a certain number of target biomolecules on the sensing area regardless of the total volume of sample fluid. For a lower concentration sample, the sample fluid with a larger volume will contain enough biomolecules to cause the differential value to change to greater than 3σ. For example, the LOD can go down to 109 fM, which is a 10-times smaller concentration than the 1.09 pM given by the previous LOD estimation, with a sample volume of 150 μL (i.e., 10 times the 15 μL volume used in the previous analysis). The same analysis can be applied to the LOD for the CRP detection. For the smaller sensing area of 0.02 mm^2, the LOD for CRP can be improved to 3.02 pM from 377 pM with the sensing area of 2.5 mm^2. With improved functionalization and passivation processes, assuming only 20% of target molecules will be lost, the LOD for CRP will be improved further to 302 fM. Finally, with the sample volume of 150 μL, the LOD can reach up to 30.2 fM.

The design presented in the work was optimized for a silver thickness of 20 nm. It is possible to design an optical cavity structure with a better LOD with a thicker silver layer. A thicker silver layer will increase the reflectance of the partially reflective mirrors which will, in turn, increase the quality (Q) factor of the optical cavity (i.e., a sharper resonance curve). With a sharp resonant response, the intensity changes of two wavelengths will become steeper, as the sensing layer's thickness increases, than those with the current design. This will enhance the differential value change and, therefore, improve the LOD. However, a thicker silver layer will also increase the absorption of light, increasing the optical loss of two laser diodes. The optical loss at each partially reflective mirror will reduce the sharpness of the resonant response. Therefore, an improved optical cavity design with a thicker silver is possible, but it has to be experimentally optimized due to these conflicting phenomena.

5. Conclusions

The optical cavity-based biosensor (OCB) has been developed for the purpose of POC biosensing. It is a label-free system detecting the local refractive index change due to the adsorption of target biomolecules on the receptor molecules. It is a low-cost system with simple and straightforward fabrication processes and low-cost parts and components, which achieves high sensitivity by employing the differential detection method. To demonstrate the limit of detection (LOD) of the OCB experimentally, we conducted streptavidin and CRP detection tests. For a silver thickness of 20 nm, the optimized optical cavity structure has a cavity height of 8 μm and a SOG thickness of 400 nm for the wavelengths of 830 nm and 904 nm. The fabricated devices have typical layer thicknesses of 22 nm (silver), 410 nm (SOG), 6.4 μm (SU8), and 1.08 μm (UV glue). The SOG surface was functionalized by the vapor-phase deposition of APTES followed by sulfo-NHS-biotin covalent bonding for more reproducible and stable test results. The polymer swelling property was used to fine-tune the optical cavity width. From the triplicate test results for streptavidin detection, the LOD of the OCB was determined to be 1.35 nM with four different concentrations of streptavidin. Human CRP was chosen to demonstrate our OCB's ability to detect an actual biomarker for the first time. With biotin-conjugated CRP antibodies as the receptor molecules, the OCB successfully detected CRP with an LOD of 377 pM. All measurements were done using a small sample volume of 15 μL in a short time, less than 30 min, once the optical cavity reached the measurable condition after the fine-tuning process. We showed that the LOD of our OCB can be improved further into the femto-molar range for both streptavidin and CRP by reducing the sensing area, improving the functionalization and passivation processes, and increasing the sample volume. The LOD of the OCB could be possibly improved with a thicker silver layer, but it must be experimentally optimized.

Author Contributions: Conceptualization, S.K.; methodology, D.R.; software, D.R.; validation, D.R.; formal analysis, D.R.; investigation, D.R. and S.K.; resources, S.K.; data curation, D.R.;

writing—original draft preparation, D.R.; writing—review and editing, S.K.; visualization, D.R.; supervision, S.K.; project administration, S.K.; funding acquisition, S.K. All authors have read and agreed to the published version of the manuscript.

Funding: This research was funded by National Science Foundation (NSF), grant number CBET-1706472 and ECCS-1707049.

Institutional Review Board Statement: Not applicable.

Data Availability Statement: The data presented in this study are available on request from the corresponding author.

Conflicts of Interest: The authors declare no conflict of interest.

References

1. Quesada-González, D.; Merkoçi, A. Nanomaterial-based devices for point-of-care diagnostic applications. *Chem. Soc. Rev.* **2018**, *47*, 4697–4709. [CrossRef]
2. Noah, N.M.; Ndangili, P.M. Current Trends of Nanobiosensors for Point-of-Care Diagnostics. *J. Anal. Methods Chem.* **2019**, *2019*, 2179718. [CrossRef] [PubMed]
3. Tobore, T.O. On the need for the development of a cancer early detection, diagnostic, prognosis, and treatment response system. *Future Sci.* **2020**, *6*, FSO439. [CrossRef] [PubMed]
4. Qian, L.; Li, Q.; Baryeh, K.; Qiu, W.; Li, K.; Zhang, J.; Yu, Q.; Xu, D.; Liu, W.; Brand, R.E.; et al. Biosensors for early diagnosis of pancreatic cancer: A review. *Transl. Res.* **2019**, *213*, 67–89. [CrossRef] [PubMed]
5. Salek-Maghsoudi, A.; Vakhshiteh, F.; Torabi, R.; Hassani, S.; Ganjali, M.R.; Norouzi, P.; Hosseini, M.; Abdollahi, M. Recent advances in biosensor technology in assessment of early diabetes biomarkers. *Biosens. Bioelectron.* **2018**, *99*, 122–135. [CrossRef]
6. Andryukov, B.G.; Besednova, N.N.; Romashko, R.V.; Zaporozhets, T.S.; Efimov, T.A. Label-Free Biosensors for Laboratory-Based Diagnostics of Infections: Current Achievements and New Trends. *Biosensors* **2020**, *10*, 11. [CrossRef]
7. Inan, H.; Poyraz, M.; Inci, F.; Lifson, M.A.; Baday, M.; Cunningham, B.T.; Demirci, U. Photonic crystals: Emerging biosensors and their promise for point-of-care applications. *Chem. Soc. Rev.* **2017**, *46*, 366–388. [CrossRef]
8. Sin, M.L.; Mach, K.E.; Wong, P.K.; Liao, J.C. Advances and challenges in biosensor-based diagnosis of infectious diseases. *Expert Rev. Mol. Diagn.* **2014**, *14*, 225–244. [CrossRef]
9. Khansili, N.; Rattu, G.; Krishna, P.M. Label-free optical biosensors for food and biological sensor applications. *Sens. Actuators B Chem.* **2018**, *265*, 35–49. [CrossRef]
10. Pirzada, M.; Altintas, Z. Recent Progress in Optical Sensors for Biomedical Diagnostics. *Micromachines* **2020**, *11*, 356. [CrossRef]
11. Gauglitz, G. Critical assessment of relevant methods in the field of biosensors with direct optical detection based on fibers and waveguides using plasmonic, resonance, and interference effects. *Anal. Bioanal. Chem.* **2020**, *412*, 3317–3349. [CrossRef] [PubMed]
12. Pathak, A.K.; Singh, V.K.; Ghosh, S.; Rahman, B.M.A. Investigation of a SPR based refractive index sensor using a single mode fiber with a large D shaped microfluidic channel. *OSA Contin.* **2019**, *2*, 3008–3018. [CrossRef]
13. Sharma, S.; Zapatero-Rodriguez, J.; Estrela, P.; O'Kennedy, R. Point-of-Care Diagnostics in Low Resource Settings: Present Status and Future Role of Microfluidics. *Biosensors* **2015**, *5*, 577–601. [CrossRef] [PubMed]
14. Hayes, B.; Murphy, C.; Crawley, A.; O'Kennedy, R. Developments in Point-of-Care Diagnostic Technology for Cancer Detection. *Diagnostics* **2018**, *8*, 39. [CrossRef]
15. Prasad, A.; Choi, J.; Jia, Z.; Park, S.; Gartia, M.R. Nanohole array plasmonic biosensors: Emerging point-of-care applications. *Biosens. Bioelectron.* **2019**, *130*, 185–203. [CrossRef]
16. Choi, J.R. Development of Point-of-Care Biosensors for COVID-19. *Front. Chem.* **2020**, *8*, 517. [CrossRef]
17. Land, K.J.; Boeras, D.I.; Chen, X.S.; Ramsay, A.R.; Peeling, R.W. REASSURED diagnostics to inform disease control strategies, strengthen health systems and improve patient outcomes. *Nat. Microbiol.* **2019**, *4*, 46–54. [CrossRef]
18. Hsieh, H.V.; Dantzler, J.L.; Weigl, B.H. Analytical Tools to Improve Optimization Procedures for Lateral Flow Assays. *Diagnostics* **2017**, *7*, 29. [CrossRef]
19. Koczula, K.M.; Gallotta, A. Lateral flow assays. *Essays Biochem.* **2016**, *60*, 111–120. [CrossRef]
20. Cho, S.; Brake, J.H.; Joy, C.; Kim, S. Refractive index measurement using an optical cavity based biosensor with a differential detection. In Proceedings of the Optical Diagnostics and Sensing XV: Toward Point-of-Care Diagnostics, San Francisco, CA, USA, 9–12 February 2015; p. 93320Z.
21. Rho, D.; Cho, S.; Joy, C.; Kim, S. Demonstration of an optical cavity sensor with a differential detection method by refractive index measurements. In Proceedings of the 2015 Texas Symposium on Wireless and Microwave Circuits and Systems (WMCS), Waco, TX, USA, 23–24 April 2015; pp. 1–4.
22. Cowles, P.; Joy, C.; Bujana, A.; Rho, D.; Kim, S. Preliminary measurement results of biotinylated BSA detection of a low cost optical cavity based biosensor using differential detection. In Proceedings of the Frontiers in Biological Detection: From Nanosensors to Systems VIII, San Francisco, CA, USA, 14–15 February 2016; p. 972509.
23. Joy, C.; Kim, S. Benefits of a scaled differential calculation method for use in a Fabry-Perot based optical cavity biosensor. In Proceedings of the 2017 Texas Symposium on Wireless and Microwave Circuits and Systems (WMCS), Waco, TX, USA, 30–31 March 2017; pp. 1–4.

24. Rho, D.; Kim, S. Large dynamic range optical cavity based sensor using a low cost three-laser system. In Proceedings of the 2017 39th Annual International Conference of the IEEE Engineering in Medicine and Biology Society (EMBC), Jeju Island, Korea, 11–15 July 2017; pp. 1393–1396.
25. Rho, D.; Kim, S. Low-cost optical cavity based sensor with a large dynamic range. *Opt. Express* **2017**, *25*, 11244–11253. [CrossRef]
26. Rho, D.; Kim, S. Label-free real-time detection of biotinylated bovine serum albumin using a low-cost optical cavity-based biosensor. *Opt. Express* **2018**, *26*, 18982–18989. [CrossRef]
27. Rho, D.; Breaux, C.; Kim, S. Demonstration of a Low-Cost and Portable Optical Cavity-Based Sensor through Refractive Index Measurements. *Sensors* **2019**, *19*, 2193. [CrossRef]
28. Purr, F.; Bassu, M.; Lowe, R.; Thürmann, B.; Dietzel, A.; Burg, T. Asymmetric nanofluidic grating detector for differential refractive index measurement and biosensing. *Lab Chip* **2017**, *17*, 4265–4272. [CrossRef]
29. Mani, R.J.; Dye, R.G.; Snider, T.A.; Wang, S.; Clinkenbeard, K.D. Bi-cell surface plasmon resonance detection of aptamer mediated thrombin capture in serum. *Biosens. Bioelectron.* **2011**, *26*, 4832–4836. [CrossRef]
30. Ozkumur, E.; Yalcin, A.; Cretich, M.; Lopez, C.A.; Bergstein, D.A.; Goldberg, B.B.; Chiari, M.; Unlu, M.S. Quantification of DNA and protein adsorption by optical phase shift. *Biosens. Bioelectron.* **2009**, *25*, 167–172. [CrossRef] [PubMed]
31. Hamed, A.M. Image processing of corona virus using interferometry. *Opt. Photonics J.* **2016**, *6*, 75. [CrossRef]
32. Ma, L.; Zhu, S.; Tian, Y.; Zhang, W.; Wang, S.; Chen, C.; Wu, L.; Yan, X. Label-Free Analysis of Single Viruses with a Resolution Comparable to That of Electron Microscopy and the Throughput of Flow Cytometry. *Angew. Chem. Int. Ed.* **2016**, *55*, 10239–10243. [CrossRef]
33. Pang, Y.; Song, H.; Cheng, W. Using optical trap to measure the refractive index of a single animal virus in culture fluid with high precision. *Biomed. Opt. Express* **2016**, *7*, 1672–1689. [CrossRef] [PubMed]
34. Zhu, M.; Lerum, M.Z.; Chen, W. How to prepare reproducible, homogeneous, and hydrolytically stable aminosilane-derived layers on silica. *Langmuir* **2012**, *28*, 416–423. [CrossRef] [PubMed]
35. Wouters, K.; Puers, R. Diffusing and swelling in SU-8: Insight in material properties and processing. *J. Micromech. Microeng.* **2010**, *20*, 095013. [CrossRef]
36. Liu, C.; Liu, Y.; Sokuler, M.; Fell, D.; Keller, S.; Boisen, A.; Butt, H.-J.; Auernhammer, G.K.; Bonaccurso, E. Diffusion of water into SU-8 microcantilevers. *Phys. Chem. Chem. Phys.* **2010**, *12*, 10577–10583. [CrossRef] [PubMed]
37. Hill, G.; Melamud, R.; Declercq, F.; Davenport, A.; Chan, I.; Hartwell, P.; Pruitt, B. SU-8 MEMS Fabry-Perot pressure sensor. *Sens. Actuators A Phys.* **2007**, *138*, 52–62. [CrossRef]
38. Fan, X. *Advanced Photonic Structures for Biological and Chemical Detection*; Springer: New York, NY, USA, 2009.
39. Qureshi, A.; Gurbuz, Y.; Niazi, J.H. Biosensors for cardiac biomarkers detection: A review. *Sens. Actuators B Chem.* **2012**, *171–172*, 62–76. [CrossRef]
40. Yang, Y.N.; Lin, H.I.; Wang, J.H.; Shiesh, S.C.; Lee, G.B. An integrated microfluidic system for C-reactive protein measurement. *Biosens. Bioelectron.* **2009**, *24*, 3091–3096. [CrossRef] [PubMed]
41. Fakanya, W.M.; Tothill, I.E. Detection of the inflammation biomarker C-reactive protein in serum samples: Towards an optimal biosensor formula. *Biosensors* **2014**, *4*, 340–357. [CrossRef]

Article

Noninvasive Monitoring of Glucose Using Near-Infrared Reflection Spectroscopy of Skin—Constraints and Effective Novel Strategy in Multivariate Calibration

H. Michael Heise [1,*], Sven Delbeck [1] and Ralf Marbach [2]

1. Interdisciplinary Center for Life Sciences, South-Westphalia University of Applied Sciences, Frauenstuhlweg 31, 58644 Iserlohn, Germany; Delbeck.Sven@fh-swf.de
2. CLAAS Selbstfahrende Erntemaschinen, Muehlenwinkel 1, 33428 Harsewinkel, Germany; Ralf.Marbach@claas.com
* Correspondence: heise.h@fh-swf.de; Tel.: +49-2371-566412

Abstract: For many years, successful noninvasive blood glucose monitoring assays have been announced, among which near-infrared (NIR) spectroscopy of skin is a promising analytical method. Owing to the tiny absorption bands of the glucose buried among a dominating variable spectral background, multivariate calibration is required to achieve applicability for blood glucose self-monitoring. The most useful spectral range with important analyte fingerprint signatures is the NIR spectral interval containing combination and overtone vibration band regions. A strategy called science-based calibration (SBC) has been developed that relies on a priori information of the glucose signal ("response spectrum") and the spectral noise, i.e., estimates of the variance of a sample population with negligible glucose dynamics. For the SBC method using transcutaneous reflection skin spectra, the response spectrum requires scaling due to the wavelength-dependent photon penetration depth, as obtained by Monte Carlo simulations of photon migration based on estimates of optical tissue constants. Results for tissue glucose concentrations are presented using lip NIR-spectra of a type-1 diabetic subject recorded under modified oral glucose tolerance test (OGTT) conditions. The results from the SBC method are extremely promising, as statistical calibrations show limitations under the conditions of ill-posed equation systems as experienced for tissue measurements. The temporal profile differences between the glucose concentration in blood and skin tissue were discussed in detail but needed to be further evaluated.

Keywords: noninvasive glucose sensing; near-infrared spectroscopy; skin tissue reflection spectroscopy; calibration modeling; science-based calibration (SBC)

Citation: Heise, H.M.; Delbeck, S.; Marbach, R. Noninvasive Monitoring of Glucose Using Near-Infrared Reflection Spectroscopy of Skin—Constraints and Effective Novel Strategy in Multivariate Calibration. *Biosensors* **2021**, *11*, 64. https://doi.org/10.3390/bios11030064

Received: 10 January 2021
Accepted: 23 February 2021
Published: 27 February 2021

Publisher's Note: MDPI stays neutral with regard to jurisdictional claims in published maps and institutional affiliations.

1. Introduction

The advantages of tight glycemic control in people with diabetes mellitus have often been documented since the diabetes control and complications trial (DCCT) studies were completed [1–3]. Those studies proved that intensive insulin therapy in such patients could dramatically delay many serious complications caused by an increase of glycation of body proteins due to above-average blood glucose concentration, which can also be connected to problems from micro- and macrovascular diseases leading, e.g., to retino-, nephro-, and neuropathy. Most diabetic patients use the equipment for blood glucose self-monitoring (SMBG) that tracks their glucose concentrations and enables them to adjust their insulin dosage and achieve normoglycemia. Over the past few years, substantial progress can be seen in research to find improved devices for diabetic patients, mostly based on electrochemical and optical sensors; for an overview on current methods and instrumentation, see recent reviews [4–11]. When undergoing intensive insulin therapy, current surveillance still requires people with diabetes to use lancets to prick their fingertips for blood sampling several times a day. Alternatively, they can use continuous glucose monitoring (CGM) devices that have recently been brought to the market. As so-called non-adjunctive devices, these still require invasive blood testing from time-to-time. Such factory-calibrated sensors are used for intermittently scanned continuous glucose monitoring but still face limitations. Problems may occur in situations with rapid blood glucose changes [12,13], and sensor glue can cause skin irritations [14]. Despite this, minimal-invasive continuous glucose sensing systems have been suggested for glycemic control in people with diabetes mellitus and critically ill patients [15].

A noninvasive measurement system certainly eliminates the inconvenience and pain of multiple daily blood tests and, as observed with continuously monitoring devices, avoids the invasiveness of today's flash glucose monitoring sensors or microdialysis catheter implants combined with ex vivo detection. Noninvasive instrumentation also allows a larger number of measurements than using SMBG invasive devices.

A multitude of optical methods has been suggested for the development of noninvasive methods of blood glucose monitoring. To date, applied spectroscopic methods have been based on vibrational spectroscopy and include mid-infrared, NIR and Raman spectroscopy, among other techniques such as fluorescence, polarimetry, and optical coherence tomography, for which recent comprehensive reviews exist, e.g., [8–11]. New publications based on mid-infrared [16] or Raman spectroscopy [17] have shown promising results for achieving noninvasive assays, and earlier papers from both research teams provide more insight into their measurement techniques [18,19].

For many years, NIR spectroscopy has found application in clinical chemistry. In particular, glucose quantification in serum, plasma, or whole blood has been demonstrated successfully by several authors; see, for example, [20–23]. Therefore, several projects were started for the development of noninvasive assays based on skin measurements. An interesting and fascinating book to read on the many fruitless efforts in the past has been published by Smith [24]. Despite those failures, NIR spectroscopy offers a substantial potential for medical applications, including noninvasive methodology for blood glucose determinations; for an overview, see a recent book chapter [25]. Skin spectroscopy based on transmission measurements requires thin skin folds or short-wave NIR spectroscopy for transilluminating a fingertip or an earlobe [26]. For accessing information-rich spectral intervals with first overtone and combination band vibrations, the diffuse reflection technique has been favored when measuring skin.

The noninvasive sensing of glucose is experimentally challenging due to the tiny glucose absorbance, a dominating high and variable background absorption of water, baseline shifts, instrumental drift, lack of sensitivity, and poor precision. Multivariate calibrations are required to allow for the obligatory selectivity [27] of reliable glucose quantification with large-enough analyte absorbances above the noise level. Traditionally, two different calibration modeling approaches are used. Analytical spectroscopists were analyzing sample spectra by least-squares fitting with reference absorptivity spectra of analytes contributing—in most cases—linearly to the sample spectrum, dependent on their concentrations. The physics behind this is the validity of Beer's law. This approach for glucose sensing is known as "classical least squares" (CLS) calibration [28] and was suggested by Maruo and Yamada [29] under the assumption that absorbance difference spectra of human forearm skin versus that of a start spectrum can be modeled by a linear combination of spectra of glucose, water, protein, fat, and a baseline for scattering.

The other more widely applied modeling technique relies on statistical calibration (also called inverse calibration) based on traditional partial least-squares (PLS), principal component regression (PCR), or, more recently, machine learning tools; for recently published examples of use in noninvasive methodology testing, see Refs [16,17]. The reader is also referred to our earlier publications for insight into previously favored calibration modeling and its advantages and disadvantages [28,30]. For proving the required selectivity, the net analyte signal (NAS) has been suggested as an approach to validate the spectrometric model when separating the glucose spectral signature from those of the tissue matrix components [31,32].

Most projects for the development of noninvasive glucose assays, including our own work, used statistical PLS calibrations with sophisticated data treatment and variable selection methods [33–36]. It could also be demonstrated that problems and pitfalls can arise from overfitting due to the implementation of too many spectral variables or insufficient model validation [33]. If a statistical calibration technique is used, there are additional problems such as a nonspecific response or an implementation of spuriously correlated spectral variance into the calibration model. Further evaluations of such spectrometric assays, including a discussion of problems and perspectives, have been published in the past [31,33].

Another approach, originally called "spectral Wiener filtering" and known from time signal processing theory, has been developed and successfully tested. The results of this approach are presented here, combining a priori information such as the spectral absorptivities of the analyte of interest with estimates of the variance of the population with negligible analyte concentration dynamics [37–39]. When compared to calibration modeling based on PLS and with regard to workload, this method is also less expensive allows, without an analytical reference method, the specificity of response to be proven from first principles, and combines the best features of both worlds, i.e., of the physical and statistical modeling approaches. Early users from the pharmaceutical industry, working with process analytical technologies (PAT), referred to this as a "science-based" method, so the name "science-based calibration" (SBC) method was created. The calibration method has been implemented several times for pharmaceutical applications, such as in a tablet-coating process using Raman spectra [40].

In the context of clinical chemistry for glucose quantification, this method was successfully tested on the NIR-spectra of plasma samples obtained using a thermostated cell with a constant pathlength of 1 mm. Compared to previous PLS calibration models, the results were favorable [33]. For transcutaneous spectra obtained by diffuse reflection, estimating the glucose "response spectrum" is more difficult since the wavelength-dependent photon penetration depth into the skin requires a wavelength-dependent scaling of the aqueous glucose absorptivity spectrum. The scaling can be obtained from optical skin parameters such as absorption and scattering constants. Results for glucose concentration in the lip mucosa tissue of a diabetic subject, recorded under modified oral glucose tolerance test conditions, will be presented using scaling parameters as obtained for dermis and lip spectra. For the first time, results are shown for tissue glucose concentrations that differ, as expected, from blood glucose measurement as the current gold standard for diabetes therapy. Several publications addressed the time delay observed in measurements within the interstitial tissue compartment as accessible with invasive needle-type biosensors. However, the present results provide insight into integral tissue measurements with vascular, interstitial, and intracellular glucose-containing aqueous compartments.

2. Materials and Methods

2.1. Spectrometer Hardware and Recorded Spectra

Within the so-called "therapeutic window"—as coined by biospectroscopists—with wavenumbers between 16,600 cm^{-1} and 7700 cm^{-1}, a transmission measurement, e.g., of a fingertip, is feasible with tissue absorption small compared to scattering [26]. For shorter wavenumbers, such NIR-spectral measurements lead to extremely low transmittance values, whereas experiments using diffuse reflectance (DR) can easily be realized. Different accessories were employed based on either quartz fibers or mirror optics (see Figure 1). With fiber-optic probes, optimal distances between tissue illuminating fibers and detection fibers can be arranged to reach a certain skin depth and to probe the dermis vasculature as well (see also different fiber arrangements illustrated in Figure 1a). This approach made use of an optimized accessory with a rotational ellipsoidal mirror, which produced the highest signal-to-noise ratios, good reproducibility of the lip measurement, and allowed a temperature control of the lip contact area at 37 °C [41].

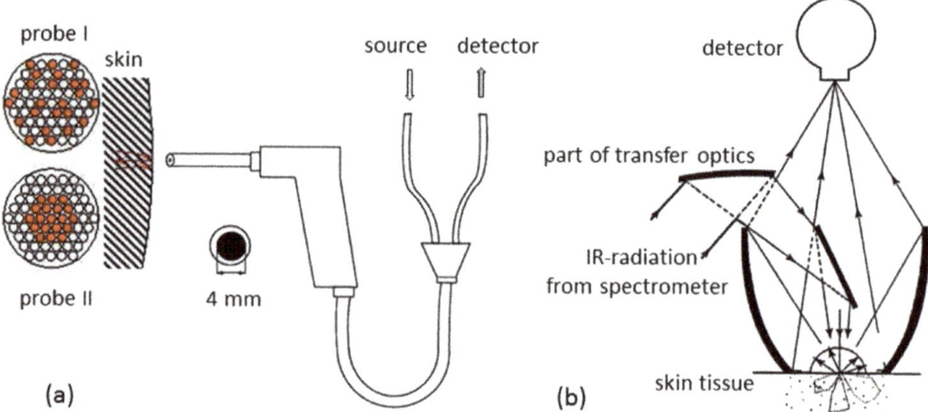

Figure 1. Diffuse reflection accessories used for skin measurements: (**a**) fiber-optic probes with fibers for illumination and detection arranged differently; (**b**) mirror-based device providing tissue spectra with different probing depths.

In Figure 2a, examples of spectra of different layers of muscle tissue backed with a reflecting gold-coated carrier substrate are shown that were measured in diffuse reflection. It shows that the spectral intensities are accessory-dependent due to the accessible solid angle for detection of the back-scattered photons. For some spectral intervals, an enlargement of the tissue thickness does not lead to larger absorbance values, indicating that "saturation" was already reached at an even shorter penetration depth than the thinnest layer of 0.8 mm would allow. For comparison, an absorbance spectrum from a 0.5 mm thick tissue layer, enclosed by quartz windows and measured in transmission, is shown together with water spectra that were measured in cuvettes of different pathlengths (see Figure 2b). In comparison with absorbance spectra of liquid water, a transmission equivalent sample thickness can be estimated for the spectra measured by diffuse reflection spectroscopy.

Mucous lip tissue was chosen for the transcutaneous measurements because it is rich in capillary blood vessels (equivalent to an advantageously high blood volume inside the probed tissue) and also provides good optical contact to the constructed diffuse reflection accessory for recording high-quality in vivo spectra. NIR-spectra of the inner lip were recorded using a Fourier Transform spectrometer (model IFS-66 from Bruker Analytische Messtechnik, Ettlingen, Germany) and a liquid nitrogen-cooled InSb detector (Infrared Associates, Suffolk, UK). An on-axis ellipsoidal mirror is the essential optics part of the reflection accessory that collects the diffusely back-reflected radiation from the skin tissue in a very efficient manner and much differently from commercially available accessories or fiber-optic probes [41,42]. Reflectance spectra $R(\tilde{v}) = I_s(\tilde{v})/I_0(\tilde{v})$ were calculated with $I_s(\tilde{v})$ and $I_0(\tilde{v})$, i.e., the single-beam lip spectrum and the reflectance standard, respectively. Previously, also spectra of standards of different reflectivity were measured, but the one with a reflectivity of 5%, also from Spectralon (Labsphere, North Sutton, NH, USA), was preferred for matching the reference standard interferogram to that of the lip spectrum in an optimal manner by comparing their interferogram maxima.

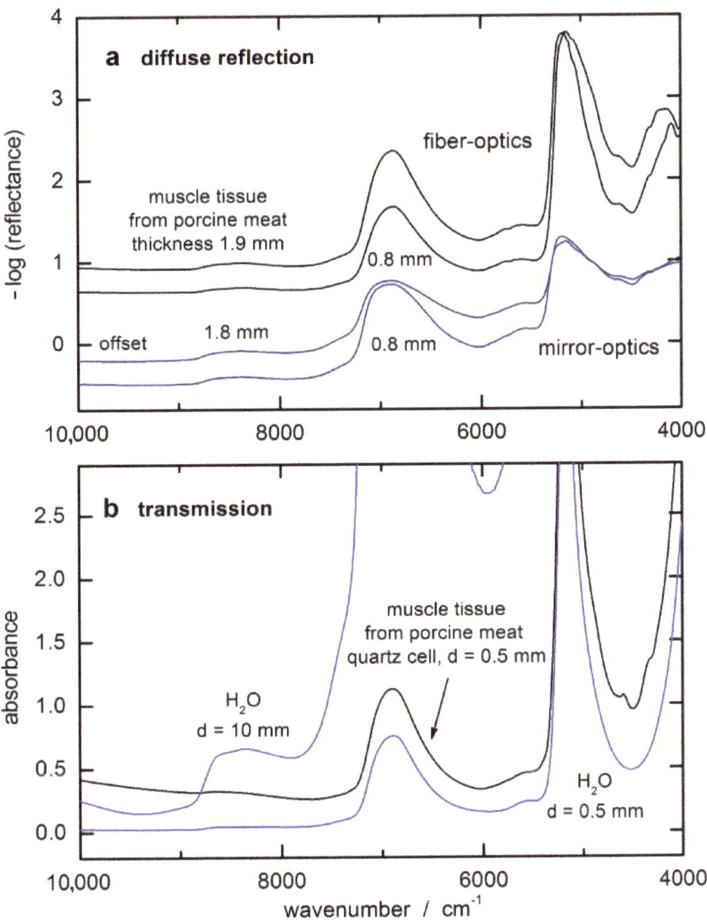

Figure 2. Spectra of tissue phantoms of constant layer thicknesses illustrating wavelength-dependent radiation penetration: (**a**) monitored with different accessories (for the first two spectra, a fiber-optic probe of type I was used); (**b**) transmission spectra of water and tissue.

Before the lip measurements were started, the test person had to rinse their mouth with plain water. After patting the lip dry, the spectra were taken reproducibly by slightly pressing the inner lower lip (oral mucosa) against the half-spherical lens of the reflection accessory. The duration of the lip measurements was 1 min for averaging 1200 single-sided interferograms providing a spectral resolution of 32 cm^{-1} after Fourier-transformation and phase correction. For examples of single-beam spectra of oral mucosa tissue and gray Spectralon standard with 5% reflectance, see also Figure 3a. The lip spectrum and noise level, obtained from two consecutive tissue measurements with baseline correction, is given in absorbance equivalent units in Figure 3b. The log-converted single-beam lip spectrum is also displayed, which shows differences only in a smooth and constant baseline compared to the −log (reflectance) spectrum. The hardware noise within the log-converted single-beam spectra is smaller by a factor of $\sqrt{2}$ than that found in the absorbance spectra as calculated from two noise-loaded single-beam measurements.

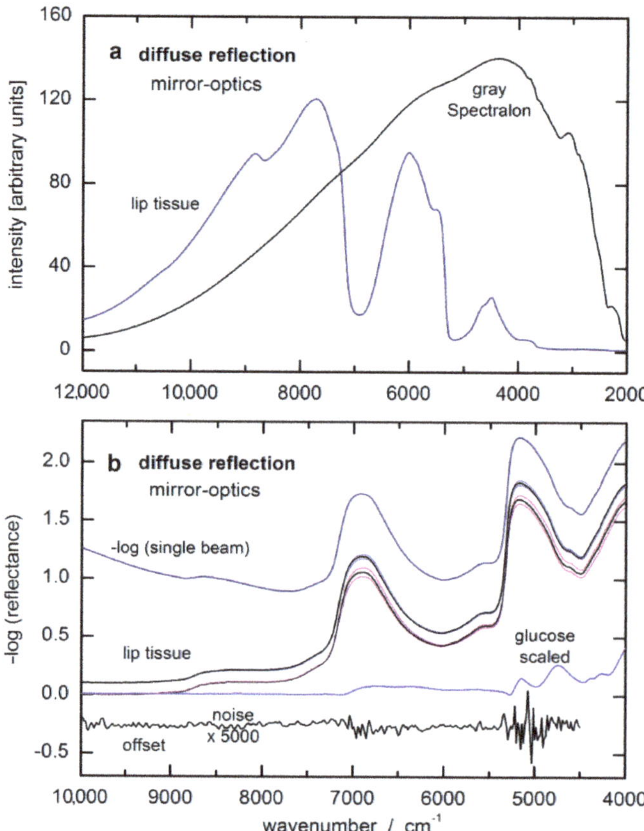

Figure 3. (**a**) Single-beam spectra of oral mucosa and a gray reflectance standard of 5% reflectivity, measured by diffuse reflection; (**b**) log-converted single-beam lip spectrum and two lip reflectance spectra as calculated from two respective series of five consecutively recorded lip tissue spectra shown as average spectra with according standard deviation spectra (spectra were taken at the beginning of our two-day measurements with capillary blood glucose values around 50 mg/dL and around 400 mg/dL during the first day; all spectra were offset for better visualization) and enlarged noise spectrum from two consecutive lip measurements after baseline correction.

2.2. Calibration Design and Reference Method

The measurements presented here were conducted as part of the initial proof-of-concept study, and although a description of earlier outcomes has been published in previous work [34,35], the dataset is nonetheless well-suited to provide additional insight into various calibration strategies and issues involved. The experiments were in compliance with ethical principles for human studies and required the consent of the informed subject. Spectra of the lip mucosa of a male test person with type 1 diabetes mellitus were recorded under the conditions of a modified oral glucose tolerance test (OGTT) over a two-day trial with parallel measurements of capillary blood glucose concentrations.

The OGTT of the first day started in the morning at a low blood glucose concentration after fasting. For reaching an increased blood glucose level after 90 min, a potion with 50 g of glucose (Dextro OGT by Boehringer Mannheim, Germany) was ingested 20 min after the initial blood glucose testing. Another load of carbohydrates was given so that when the maximum blood glucose concentration was reached, an appropriate dose of insulin was injected to achieve a steady reduction in glucose concentration within another 90 min.

The test carried out during the second day started with a slightly larger than above fasting level of blood glucose concentration of the subject. After the lapse of 90 min, the glucose amount of 50 g was taken with some liquid. The total duration for the OGTT experiments was 15 h for collecting our total 133 lip spectra.

Capillary blood glucose reference values were measured using a clinical chemistry-established enzymatic assay (D-glucose, Boehringer Mannheim, Germany) based on both enzymes of hexokinase (HK) and glucose-6-phosphate dehydrogenase (G6PDH), as programmed on the analytical instrument model ACP 5040 (Eppendorf, Hamburg, Germany). The coefficient of variation for the reference measurements was less than 4%, as validated by control sera, which is more accurate compared to many studies using test strip devices with a 15% span of relative uncertainty. Over the two days, capillary blood was sampled with 20 μL capillary pipettes from Brand (Wertheim, Germany) after pricking the fingertip and subsequent recording of the exact sampling times. Blood glucose reference readings obtained during this two-day experiment ranged from 30 mg/dL (1.7 mmol/L) to 600 mg/dL (33.3 mmol/L). The population mean value (c_{av}) calculated from these measured values was 301 mg/dL (16.7 mmol/L) with a standard deviation of SD = 168 mg/dL (9.3 mmol/L). As the times of blood sampling for our reference measurements did not match those of the collected spectra (the time gap between two reference measurements was approximately 15 min), interpolation between the capillary blood measurements was carried out by spline approximation for calculating the blood concentration values at the time of the spectral lip measurements.

2.3. Optical Data for In Vivo Glucose Calibrations

For elucidating probed tissue volumes and photon penetration depths, the photon fluence rate in turbid media can be estimated by different mathematical tools for calculating the radiative transfer. The basis for such calculations are the tissue-optical properties, i.e., absorption and the scattering coefficients, μ_a and μ_s (in units of mm^{-1}), respectively, and the anisotropy of scattering g (dimensionless). Using the latter two parameters, also the reduced scattering coefficient $\mu'_s = \mu_s (1 - g)$ can be calculated. From simple diffusion theory, the optical penetration depth can be estimated as $\delta = (3 \mu_a (\mu_a + \mu'_s))^{-1/2}$. For more information, the reader is referred to a tutorial on this subject [43]. As evident from the wavelength-dependent optical tissue constants, the average optical pathlength for radiation within mucosa tissue is wavenumber-dependent, as explicated above. Results from Monte Carlo simulations of the radiation transport, simulating the "photon random walk", were presented in the past for the above-mentioned reflection accessory [41].

In Figure 4, optical constants from skin measurements are compiled from three recent publications [44–46]. Besides absorption and scattering coefficients, also the anisotropy factor g is shown in the inset of Figure 4b, which is responsible for the main forward scattering characteristics of photons within the NIR spectral range. The optical data for the epidermis and dermis from Salomatina et al. [45] are different from the other compilations, which is certainly understandable for the thin epidermis layer.

In order to unambiguously detect glucose throughout a complex matrix of several substances within the measured absorption spectrum, a sufficiently large contribution of the analyte signal to the overall spectrum is required that is significantly different from and above the experimentally observed spectral noise. Figure 5 shows the optical constants of glucose obtained from spectral measurements of aqueous solutions in the laboratory. These are of special importance in the field of in vivo NIR spectrometry since it is predominantly the analyte absorption stemming from glucose in the aqueous tissue compartments, i.e., the vascular space and interstitium, that are of interest. The spectra were obtained by scaled subtraction of a pure water absorbance spectrum. The dips in the resulting spectra can be explained by insufficient water compensation attributable to differences in the hydrogen bonding network in pure water and in glucose solutions, respectively. It must be noted that the features of glassy sugar, as-received from syrup preparations after water removal by careful heating, show the same wavenumber dependency as the aqueous solution phase spectrum. Using this technique, the solution-opaque spectral ranges due to large water absorptivities are now accessible, despite still uncompensated water absorption features within the intervals of intense absorption bands. Spectra of crystalline glucose, as well as glucose monohydrate, are also shown, which were scaled for comparison.

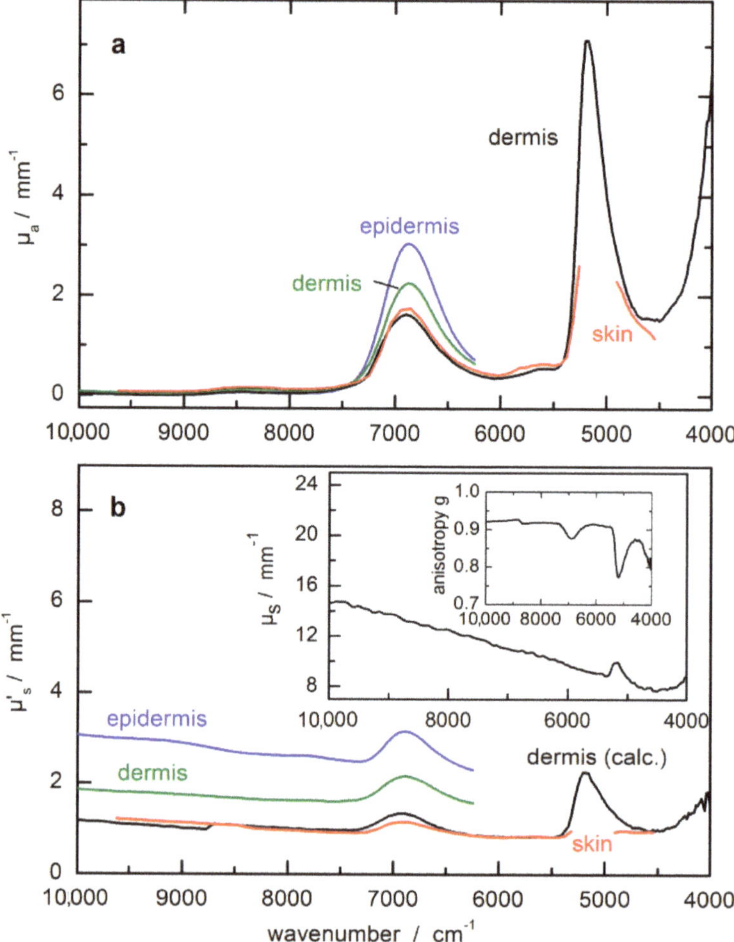

Figure 4. Optical constants of skin, dermis and epidermis from NIR-spectral measurements. (a) $\mu_a(\tilde{v})$ = absorption coefficient; (b) $\mu'_s(\tilde{v})$ = reduced scattering coefficient with $\mu'_s(\tilde{v}) = (1 - g(\tilde{v}))\,\mu_s(\tilde{v})$; this parameter includes the scattering coefficient $\mu_s(\tilde{v})$ and the anisotropy factor $g(\tilde{v})$; optical constants were compiled from refs. [44–46]; obvious differences in the scattering coefficients within the near-infrared (NIR)-spectral range must be noted.

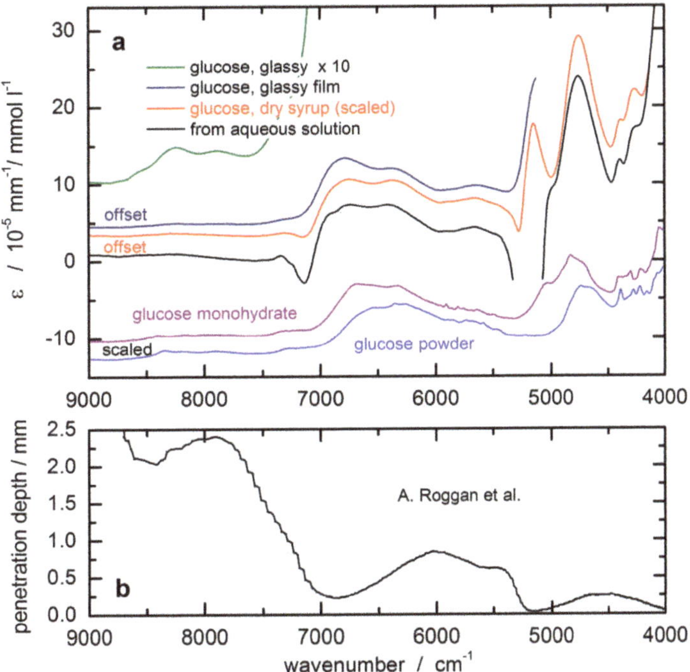

Figure 5. (**a**) Absorptivities of glucose calculated from transmission measurements of aqueous solutions and of glass-like sugars from dehydrated syrup sample (scaled to the aqueous solution phase absorptivities), as well as from glucose powder; (**b**) optical penetration depth based on radiation diffusion theory (from Roggan et al. [46]) used for scaling the glucose response spectrum for the noninvasive assay.

3. Chemometrics Based on SBC

As the level of familiarity with this method is rather low, a short outline and description of the mathematics is allowed. For the SBC method, the spectral analyte signal is estimated from a physical point of view and the spectral noise by using statistical tools. This can combine the accuracy of an inverse model with relatively low calibration effort and the simplicity of interpretation of a physics-based classical approach. The computational effort of calibration can be considerably diminished with respect to current routine practice using statistical calibrations since the requirement for a large population of calibration samples is no longer necessary. The previously intangible attribute of the analyte's response specificity is thus based on spectroscopic first-principles, eliminating the need for analytical reference methods for calibration standards, which are essential for PLS calibrations. For computer implementation, the SBC software package was programmed in MATLAB (MathWorks, South Natick, MA, USA).

Theory and Background

The following NIR spectra are given in units of [AU], from which the analyte concentrations are determined. Here, the analyte of interest is glucose (given in mg/dL). If all spectroscopic factors contributing to the spectrum are available, the experimental NIR spectrum can be described with the following equation:

$$\mathbf{x}^{T}(t) = y(t) \cdot \mathbf{g}^{T} + c_1(t) \cdot \mathbf{k}_1^{T} + c_2(t) \cdot \mathbf{k}_2^{T} + \ldots + \mathbf{i}_{baseline}^{T}(t) + \ldots + \mathbf{i}_{noise}^{T}(t) \qquad (1)$$

where the vector $x^T(t)$ is the experimental spectrum (the transpose sign "T" means that the spectrum is written as a row vector). This vector $x^T(t)$, as well as its compounds, are time-dependent functions of (t). The true, and here sought after, glucose concentration is given as a scalar and described as the "analyte concentration" $y(t)$. The "analyte response spectrum" is g^T with units of (AU/(mg/dL)); the concentrations $(c_1(t), c_2(t), \ldots c_n(t))$ and respective response spectra $(k_1^T, k_2^T, \ldots k_n^T)$ contain all information on spectral perturbations that can be explained by tissue components (i.e., water, proteins such as collagen and albumin, blood and interstitial fluid components, and others). The spectra $i_{baseline}^T(t), \ldots, i_{noise}^T(t)$ include all influencing factors that are produced by the spectrometer and its sampling interface, such as, but not limited to, detector noise, baseline slopes, and shifts, for example, from scattering differences, etc. As the SBC method summarizes all "non-glucose-related" factors in a single expression, Equation (1) can be shortened to:

$$x^T(t) = y(t) \cdot g^T + x_n^T(t) \tag{2}$$

where $x_n^T(t)$ represents all factors belonging to the experimental spectrum, such as effects from instrumentation or spectra from interferents, but excludes contributions from the sought-after analyte. The first term, denoted by "$y(t) \cdot g^T$", is the "spectral signal", whereas the second term "x_n^T" will be noted as "noise".

The key issue of the SBC method is knowing the response spectrum of the analyte, g^T. However, for a noninvasive glucose measurement from diffuse reflection spectra of skin, the situation is quite different from a transmission measurement with given sample thickness as provided by a cuvette for measuring whole blood or blood plasma as in routine clinical chemistry applications. The noninvasive approach requires not only the glucose absorptivities but also the wavenumber-dependent "effective pathlength" within the probed tissue for a chosen accessory for diffuse-reflection measurements.

The spectral signal and spectral noise can be described by their first- and second-order statistics [37]. The spectral signal can be defined by a mean, $\bar{y} \cdot g^T$, and a root-mean-square (RMS) term, $\sigma_y \cdot g^T$, where g^T is the analyte response spectrum. In the case of people with type-I diabetes, the standard deviation σ_y for the varying blood glucose levels $y(t)$ can be as large as around 90 mg/dL. Spectral noise thus can be described by a vector of mean, $\bar{x}_n^T$, and a covariance matrix Σ, where the latter provides all spectral changes, which occur independently of the sought-after analyte, i.e., the variation from interferents and additional instrumental effects. It is advantageous that for noninvasive glucose measurements, the spectral noise Σ can also easily be determined by recording N spectra from healthy people with near-constant glucose concentrations reflecting the spectral tissue variations over time independent of blood glucose changes. By forming these spectra x_n^T into a matrix X, our covariance matrix Σ is calculated as

$$\Sigma \cong \widetilde{X}^T \widetilde{X}/(N-1)[AU^2] \tag{3}$$

where the tilde ("~") indicates a mean-centered matrix. If a "subject-specific" estimate of the spectral noise is required, spectra from a single test subject can be sampled over time to estimate the noise covariance. With regard to the instrument-to-instrument "noise," the determination of Σ virtually always has relatively low experimental effort since reference values of the analyte concentration are not essential, as described above. In addition, should there be any variation in the glucose levels that are present in the experimental spectra collected for estimating Σ, the calibration method will still work (for further details, we refer to ref. [33]).

By applying the notation above, the optimal regression vector (also known in literature as "b-vector") for the analyte determination is calculated by:

$$b_{opt(1)} = \Sigma^- g/(g^T \Sigma^- g) \ [(mg/dL)/AU] \tag{4}$$

with Σ^{-} being the inverse of Σ. Please note that Equation (4) provides a mean-square "prediction" error minimum under the condition of unity prediction slope, necessary for measurement purposes, as illustrated by a scatter plot of predicted versus reference concentration values (indicated in the subscript). When $\mathbf{b}_{opt(1)}$ is applied for the "prediction" of the analyte concentration from a newly measured spectrum, $\mathbf{x}^{T}_{pred}$, its concentration y_{pred} is calculated by:

$$y_{pred} = \overline{y} + \left(\mathbf{x}_{pred} - \overline{\mathbf{x}}\right)^{T} \cdot \mathbf{b}_{opt(1)} \ (mg/dL) \tag{5}$$

with $\overline{y}$ being the mean analyte concentration and $\overline{\mathbf{x}}^{T}$ being the mean noise spectrum of the individual spectra, which were employed for estimating Σ. With these definitions, the RMS prediction error, also known as the standard deviation of $(y_{pred} - y)$, is calculated by:

$$SEP_{opt} = (1/(\mathbf{g}^{T} \cdot \Sigma^{-} \cdot \mathbf{g}))^{1/2} \ (mg/dL) \tag{6}$$

If we look into the dependencies of the b-vector optimum (Equation (4)) and the detection limit (Equation (6)), neither are dependent on reference values since these only depend on spectroscopic data.

The issue of selectivity in the multivariate calibration case has often been discussed in the context of performance characteristics of analytical methods. The concept of "net analyte signal"—see, e.g., refs. [27,31,32]—is useful and approximates the correct definition in those stable measurement conditions where, after orthogonalization against all "other" components, there is still a sufficient analyte response spectrum that is well above the instrument noise floor. As noted in a previous publication, the net-analyte concept is insufficient and inconsistent for routine experience in many NIR-spectroscopic and other challenging applications with statistical calibration modeling [47]. The correct definition of selectivity is mathematically straightforward when using the SBC-scheme and nomenclature [33].

A number of important advantages of SBC are evident from the discussion above:

- Laboratory-based analytical work is made virtually needless for calibration for allocating reference values (for validation, this will often still be necessary). Thus, the workload of calibration is significantly lowered when comparing it to statistical calibration effort with PLS;
- The "noise" spectra required for an estimate of the "skin noise" can be sampled from healthy test subjects instead of from people with *diabetes mellitus*. The fact that these "normal subjects" will show only a narrow glucose variance is certainly an SBC advantage;
- Selectivity of response can easily be proven to regulatory agencies and concerned practitioners. Method validation, which requires an application-specific assessment, also becomes easier.

Finally, we remark that the SBC "method" is not an algorithm per se. It uses Equation (4) to compute the b-vector, and for this, the user is asked to provide estimates of both important calibration parameters, i.e., the signal $\mathbf{g}$ and the noise Σ. When both estimates describe reality well, the resulting calibration is the so-called "matched filter" and achieves the globally optimal mean-square error.

4. Results and Discussion

Some preliminary information must be mentioned before presenting the results from our two-day OGTT test with a type-1 diabetic subject. For concentration prediction, we used the—log (reflectance) spectra of the inner lip as absorbance equivalent data and an SBC calibration vector calculated as follows: The calibration interval was from 8994–5477 cm^{-1} (115 data points).

First, the spectral noise, Σ, was estimated as the sum of four variance terms. As an intrinsic term, the hardware noise can be found on the diagonal matrix elements of Σ, which was estimated to be 30 μAU RMS at 6300 cm^{-1}. At other wavenumbers, this value was scaled by the inverse of the intensity of the single-beam of the (average) lip spectrum, thus becoming wavenumber-dependent. Second, offset noise—defined as a spectrally flat baseline with an amplitude randomly varying with 50 mAU RMS – was added; third, the spectral variation from the irreproducibility of the lip contact, also known as "lip-noise" covariance, was calculated from a population of differences of spectra less than 8 min apart to minimize residual glucose features. Forth, "water displacement noise" was constructed by using a pathlength scaled water spectrum (see details of the pathlength scaling below), calculated from absorptivity data from Bertie [48], and the amplitude scaled to 2% RMS of displacement. The MATLAB™ codes used in the calculation of these four variance terms were all very similar to the codes given in [33].

The "water displacement noise" was added to the estimate of Σ in order to break the unspecific correlation (UC) that exists between the glucose concentration in the skin and its water concentration. The unspecific correlation between the glucose and the water concentrations is due to displacement, i.e., the water concentration is decreasing whenever the glucose concentration is increasing and vice versa. This effect was determined to be the dominant UC effect for the in vitro measurement of glucose [33], and therefore is expected to be important also for the noninvasive case. With the covariance matrix set up in this manner, its inverse was computed at full rank.

The glucose response spectrum used for the SBC calibration was calculated from that of an aqueous solution measured by using a cuvette of 0.5 mm pathlength. It still shows negative features around 7200 cm^{-1} due to water absorbance overcompensation (see Figure 5a). For scaling the glucose absorptivity spectrum, the optical-penetration-depth spectrum, as provided by Roggan et al. [46], see Figure 5b, was used. These values were multiplied by a constant factor of 0.4 in order to account for the reduced glucose concentration in tissue compared to whole blood or blood plasma, by which the spread of experimental blood glucose concentration values was also reproduced.

The experiments with the lip measurements were carried out without an exact repositioning scheme, meaning that the position of the spectroscopically recorded lip area of ca 2 mm diameter was randomly distributed across an area of roughly one cm^2. Repositioning of the optical probe, for example, during an experiment using a rat animal model (see ref. [32]) led to a significant scatter in glucose prediction, so that the quality of our lip spectral data with regard to reproducibility and low-noise must be highly rated, especially when the second-day spectra are taken into account with the test subject showing more routine in lip measurements. Despite the temperature control of the lip contact area of the accessory, still, temperature gradients can be manifested, as evident from a principal component analysis (PCA) of the spectral population [36]. The dominating factors stem from the water spectrum and its temperature dependency, but also other features arising from methylene stretching overtones of the long-chain acyl groups found in the subcutaneous fatty tissue become visible.

In Figure 6, the time-dependent reference blood glucose concentration values are displayed together with the SBC predictions. The predicted and reference glucose concentration values had been day-wise mean-centered to adjust for a constant offset experienced here. The raw predictions were subtracted by 20 mg/dL to align with the fasting blood glucose level when starting the monitoring over the day. The offset makes sense to "high pass filter out" a noise component that was so far not included in the calibration Σ. As a possible explanation, we suppose that the pure water spectrum declared as "water displacement noise" and added to the estimate of Σ, was not quite appropriate, since there exist differences for the tissue water due to temperature and hydrogen-bonded molecules of various tissue compounds, which can be sensitively detected within the NIR spectrum. An idea of the complexity of water signatures within lip tissue can be obtained from a principal component analysis (PCA) of the lip spectra, which has been earlier illustrated [36]. At least the first five factors contributing to the spectral variance show features related to water and its dependencies on temperature gradients and differences in the hydrogen-bonded network.

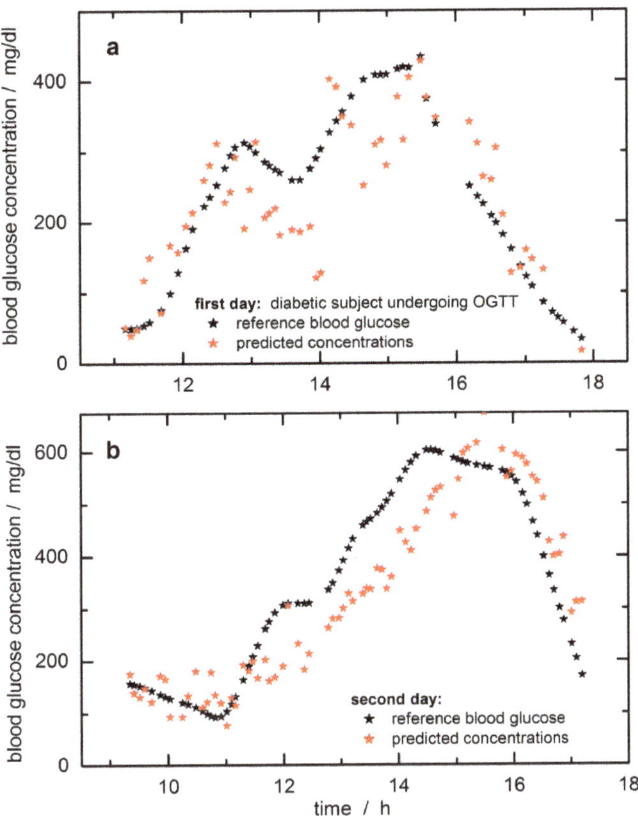

Figure 6. Predictions of tissue glucose concentrations versus blood glucose values for the first day (**a**) and the second day of the modified oral glucose tolerance test (OGTT) (**b**); tissue concentrations were calculated with the SBC-calibration model using reflection lip spectra and glucose absorptivities scaled with photon penetration depth (results were offset corrected, see also text). Spectra and reference blood glucose concentration data are from previous publications [31,33]; calibration interval was from 8994–5477 cm^{-1} (115 data points) with noise covariance calculated from differences of lip spectra less than 8 min apart.

For the first day, during the starting phase, a slightly "running ahead" of the time profile of the tissue glucose is observed compared to the reference capillary concentration values from the fingertip, whereas for other periods, a significant time lag can be noticed. A similar lack of time correlation can be observed for the second day (this trace is also showing a reduced scatter when compared with the prediction data of the first day; note the difference in the Y-axis scales). Two days' worth of data from a single subject is not a large enough data set to allow quantitative statements about "typical" time behavior, but the following conclusions are clear. Relative to the time profile of the blood glucose concentration, the time profile in the skin can lead or lag. In several publications, it has been noticed that a glucose decrease in tissue can drop earlier than found for the vascular compartment, i.e., a glucose decrement in tissue precedes hypoglycemia [49]. On the other hand, also such a feature has been found for situations with a lead-in tissue for glucose increments (see experiment shown in [32]). The lack of time-correlation displayed by the examples in Figure 6 is concerning. The mechanisms causing this discrepancy will need to be studied in the future, and their effects quantified for given segments observed for "typical" patients in typical environments. Especially for skin inserted CGM sensors, the time delay between blood and interstitial glucose profiles has been recently studied several times with two publications given here, one with three different sensors and a second paper dealing with an in silico study [50,51]. In particular, the in silico study is extremely interesting as glucose concentration differences of up to ± 40 mg/dL between blood and interstitium were obtained, and time delays up to 25 min were realized. At any rate, even the simplest way of thinking about the glucose-in-the-skin as a time signal leads to a second-order differential equation, i.e., with two inputs, carbohydrate intake making the signal go up and injecting insulin, making it decrease. Which of these two effects wins the race to the measured skin and thus determines the slope (d/dt) of the signal at the time of measurement depends on what the subject did during the previous two hours. Given that people are a bit unpredictable even when performing routines, there cannot be an exactly fixed time-shift relationship to the glucose-in-the-blood. However, we can hope for more-or-less repeatable tendencies of patient behavior.

The SBC method also allows measuring the effect of the hardware noise floor only, i.e., by multiplying it into the b-vector for providing an estimate of the repeatability error, which was calculated to be 35.5 mg/dL RMS. For further comparison, with these data set at hand, several extensive studies have been carried out based on PLS calibrations, and the reader is referred to our publications [33,36]. Best standard errors of prediction could be achieved with variable selection, reaching a SEP = 36.6 mg/dL. Using an impulse invariant designed Butterworth filter of first-order with a time constants of 10 min, the time-dependent blood glucose profiles were shifted for allocating probably concentration reference values more similar to tissue estimates, but SEP improvements of only 2.5 mg/dL could be reached. Arnold and coworkers considered a time shift of 15 min between tissue and blood glucose profile for rat skin [32]. An offset similar as experienced within our SBC study could also be observed when using different daily data sets for calibration modeling; see results illustrated in [33]. Arnold and coworkers investigated the tissue variability and its impact on the PLS regression vectors. Differences in skin inhomogeneity led to vector changes with the effect of offsetting the prediction results.

Since there is no least-squares fitting to blood glucose values (or any other reference values) carried out in SBC calibration, the SBC predictions represent a direct measurement of the glucose concentration in the tissue. Unlike statistical calibrations, PLS, etc., which rely on the correlation between blood glucose reference values and skin spectra, and which are therefore influenced by the dynamic glucose transport processes between vascular and interstitial and intracellular compartments, SBC calibration is not influenced by these processes. It just measures the glucose in the skin. Therefore, it is rather useless to state SEP-values with systematic deviations as illustrated by in silico simulations, which yet considered two compartments only, i.e., vascular and interstitial space. The methodology applied by integral tissue spectroscopy will even show a larger complexity by taking the intracellular compartment additionally into account. To reach similar results with PLS calibrations, a large number of clamp experiments with steady-state conditions would need to be performed to furnish the analyst with the appropriate number of calibration samples and for allowing a comparison based on SEP or similar metrics used for sensor quality assessment such as MARD values (average of the absolute error between all CGM values and matched reference values). A comparison with other vibrational spectroscopy methods [16,17], recently published and mentioned in the beginning, is difficult. Results from the multiperson studies were given as MARD values of 12% for the mid-infrared spectral measurements with photothermal detection, and spectral outliers previously removed [16] or around 24% and larger for noninvasive Raman measurements [17]. For the SBC study, a MARD of 23% can be calculated with the omission of six extreme outlier data.

Since spectroscopy probes the integral tissue glucose whereas reference values used for validation rely on whole blood analysis, diffusion processes within the tissue, especially between the vascular and interstitial compartments leading to a variable temporal shift in both concentration time-series, will certainly need further investigations. It is not clear, unfortunately, whether it will be possible to predict capillary glucose concentration values uniquely from tissue measurements unless other experimental options for avoiding such complications are employed, e.g., the use of pulsatile spectroscopy (plethysmography).

5. Conclusions

The validation of calibration models for a noninvasive and transcutaneous blood glucose assay based on NIR spectrometry of skin illustrates a rather critical aspect. For the first time, the so-called "spectral Wiener filter" method, also known as a science-based method of calibration (SBC), was applied to noninvasive glucose measurement. The accuracy was tested on time-sequenced NIR-spectra collected over two OGTT days with better performance on the second day, which allowed conclusions about the behavior of the glucose concentration–time profiles in the blood and in the measured skin volume.

The accuracy achieved is far from that required for a viable noninvasive measurement. Still, the results are useful in a number of ways. First, they demonstrate that SBC calibration is possible also in the noninvasive case, i.e., that workload of calibration can be drastically diminished compared to today's routine practice with PLS usage. Second, the results indicate that also in the noninvasive case, glucose can be measured in a selective way, i.e., without using unspecific correlation effects like water displacement as a signal, which statistical calibrations have done in the past [35]. However, more work is needed here. The precision of the presented FT-NIR measurement is not good enough, i.e., the prediction scatters error is too large to allow reliable assessment of the magnitude of the prediction slope, or rather: deviation from the ideal value of unity slope. The latter is necessary for an exact, quantitative prove of selectivity of response [39]. The results shown in Figure 6 indicate a slope close to the desired value of one; for further discussion, see also ref. [33]. The accuracy of the scaling applied to transform the aqueous glucose absorptivity spectrum into the response spectrum of the noninvasive case is hard to estimate but is believed to be within several 10% of the true scaling, which is supported by the prediction of the blood glucose concentration range.

The robustness of calibration also must be improved in the future, i.e., additional variance contributions need to be added to the noise estimate Σ used in our SBC calibration (with the corresponding potential price in sensitivity) [35]. With in vivo spectroscopy, the variability from physiology, problems with repositioning of skin tissue, temperature gradients, blood flow effects, photon penetration depths, etc., are required to be further studied before calibration robustness for such application can generally be proven (e.g., avoiding systematic errors like the daily offsets experienced in our data set). The SBC prediction results also reveal that a different type of instrument will be more advantageous in the future because FT-NIR spectrometers are influenced by a relatively large hardware (detector) noise in the NIR-spectral range. In summary, the SBC calibration produced in the above way relies entirely on spectroscopy data and knowledge and does not use the laboratory reference values at all. In this regard, it is similar to the established classical calibration methods, but unlike these spectral fitting methods, the SBC result converges against, and therefore in praxis comes much closer to the optimal Wiener filter result.

The fundamental problem to be overcome in the future, however, seems to be the lack of time correlation between the glucose concentrations within the different tissue compartments, i.e., within the vascular, interstitial, and intracellular space. While the relationship of the temporal glucose profiles from the intravascular and the interstitial compartments was studied intensively by various needle-type sensors, the intracellular space also with phosphorylated glucose has so far been paid no attention apart from a few attempts. The effects causing a mismatch between the glucose concentration in blood and integral skin tissue need to be further evaluated. Though the optical spectroscopic methods have always caught much attention, and many different approaches with demanding hardware have been presented during the last 20 years, there seems to be no break-through hovering on the horizon using NIR-spectroscopy of skin tissues. On the other hand, the

SBC method can be considered a strong chemometric tool for regression and modeling tasks, and further work into noninvasive testing will be worthwhile.

Author Contributions: Conceptualization, methodology, validation, supervision: H.M.H. and R.M.; funding acquisition and project administration: H.M.H.; software development: R.M.; data analysis and visualization: S.D.; writing—original draft preparation, review, and editing: H.M.H., S.D. and R.M. All authors have read and agreed to the published version of the manuscript.

Funding: Former financial support by the German Research Foundation (DFG) is gratefully acknowledged (grant number HE 1685/1).

Institutional Review Board Statement: The study was conducted according to the guidelines of the Declaration of Helsinki, valid in 1991 when the experiments were carried out, and approved by the Review Board of the former Diabetes-Forschungsinstitut, Düsseldorf, Germany.

Informed Consent Statement: The test person had given his informed consent for inclusion before he participated in this study.

Data Availability Statement: The data presented in this study has been described in more detail in previous publications; see refs. [33–36]; the data presented were reanalyzed with novel, innovative chemometric tools.

Acknowledgments: We thank our former project partner T. Koschinsky for support and assistance in medical affairs.

Conflicts of Interest: The authors declare no conflict of interest.

Abbreviations

(AU) absorbance units; (DCCT) diabetes control and complications trial; (MARD) mean absolute relative difference; (NAS) net analyte signal; (NIR) near-infrared; (OGTT) oral glucose tolerance test; (PAT) process analytical technology; (PCA) principal component analysis; (PLS) partial least-squares; (PCR) principal component regression; (RMS) root mean square; (SBC) science-based calibration; (SEP) standard error of prediction, (SMBG) self-monitoring of blood glucose.

References

1. The Diabetes Control and Complications Trial Research Group. The effect of intensive treatment of diabetes on the development of diabetes on the development and progression of long-term complications in insulin-dependent diabetes mellitus. *N. Engl. J. Med.* **1993**, *329*, 977–986. [CrossRef]
2. The Juvenile Diabetes Research Foundation Continuous Glucose Monitoring Study Group. Continuous glucose monitoring and intensive treatment of type 1 diabetes. *N. Engl. J. Med.* **2008**, *359*, 1464–1476. [CrossRef]
3. Holman, R.R.; Paul, S.K.; Bethel, M.A.; Matthews, D.R.; Neil, H.A.W. 10-Year follow-up of intensive glucose control in type 2 diabetes. *N. Engl. J. Med.* **2008**, *359*, 1577–1589. [CrossRef] [PubMed]
4. Rodbard, D. Continuous glucose monitoring: A review of successes, challenges, and opportunities. *Diabetes Technol. Ther.* **2016**, *18* (Suppl. 2). [CrossRef] [PubMed]
5. Gonzales, W.V.; Mobashsher, A.T.; Abbosh, A. The progress of glucose monitoring—A review of invasive to minimally and non-invasive techniques, devices and sensors. *Sensors* **2019**, *19*, 800. [CrossRef]
6. Chen, C.; Zhao, X.-L.; Li, Z.-H.; Zhu, Z.-G.; Qian, S.-H.; Flewitt, A.J. Current and emerging technology for continuous glucose monitoring. *Sensors* **2017**, *17*, 182. [CrossRef]
7. Dziergowska, K.; Łabowska, M.B.; Gąsior-Głogowska, M.; Kmiecik, B.; Detyna, J. Modern noninvasive methods for monitoring glucose levels in patients: A review. *Bio-Algorithms Med-Syst.* **2019**, *15*, 20190052. [CrossRef]
8. Jernelv, I.L.; Milenko, K.; Fuglerud, S.S.; Hjelme, D.R.; Ellingsen, R.; Aksnes, A. A review of optical methods for continuous glucose monitoring. *Appl. Spectrosc. Rev.* **2018**, *54*, 543–572. [CrossRef]
9. Vahlsing, T.; Delbeck, S.; Leonhardt, S.; Heise, H.M. Noninvasive monitoring of blood glucose using color-coded photoplethysmographic images of the illuminated fingertip within the visible and near-infrared range: Opportunities and questions. *J. Diabetes Sci. Technol.* **2018**, *12*, 1169–1177. [CrossRef] [PubMed]
10. Delbeck, S.; Vahlsing, T.; Leonhardt, S.; Steiner, G.; Heise, H.M. Non-invasive monitoring of blood glucose using optical methods for skin spectroscopy—Opportunities and recent advances. *Anal. Bioanal. Chem.* **2019**, *411*, 63–77. [CrossRef] [PubMed]
11. Delbeck, S.; Heise, H.M. Evaluation of opportunities and limitations of mid-infrared skin spectroscopy for non-invasive blood glucose monitoring. *J. Diabetes Sci. Technol.* **2020**. [CrossRef]

12. Moser, O.; Eckstein, M.L.; McCarthy, O.; Deere, R.; Pitt, J.; Williams, D.M.; Hayes, J.; Harald Sourij, H.; Bain, S.C.; Bracken, R.M. Performance of the Freestyle Libre flash glucose monitoring (flash GM) system in individuals with type 1 diabetes: A secondary outcome analysis of a randomized crossover trial. *Diabetes Obes. Metab.* **2019**, *21*, 2505–2512. [CrossRef]
13. Ajjan, R.A.; Cummings, M.H.; Jennings, P.; Leelarathna, L.; Rayman, G.; Wilmot, E.G. Accuracy of flash glucose monitoring and continuous glucose monitoring technologies: Implications for clinical practice. *Diabetes Vasc. Dis. Res.* **2018**, *15*, 175–184. [CrossRef]
14. Asarani, N.A.M.; Reynolds, A.N.; Boucher, S.E.; de Bock, M.; Wheeler, B.J. Cutaneous complications with continuous or flash glucose monitoring use: Systematic review of trials and observational studies. *J. Diabetes Sci. Technol.* **2020**, *14*, 328–337. [CrossRef] [PubMed]
15. Costa, D.; Lourenço, J.; Monteiro, A.M.; Castro, B.; Oliveira, P.; Tinoco, M.C.; Fernandes, V.; Marques, O.; Gonçalves, R.; Rolanda, C. Clinical performance of flash glucose monitoring system in patients with liver cirrhosis and diabetes mellitus. *Sci. Rep.* **2020**, *10*, 7460. [CrossRef] [PubMed]
16. Lubinski, T.; Plotka, B.; Sergius Janik, S.; Canini, L.; Werner Mäntele, W. Evaluation of a novel noninvasive blood glucose monitor based on mid-infrared quantum cascade laser technology and photothermal detection. *J. Diabetes Sci. Technol.* **2021**, *15*, 6–10. [CrossRef]
17. Pleus, S.; Schauer, S.; Jendrike, N.; Zschornack, E.; Link, M.; Hepp, K.D.; Haug, C.; Freckmann, G. Proof of concept for a new Raman-based prototype for noninvasive glucose monitoring. *J. Diabetes Sci. Technol.* **2021**, *15*, 11–18. [CrossRef]
18. Pleitez, M.A.; Hertzberg, O.; Bauer, A.; Lieblein, T.; Glasmacher, M.; Tholl, H.; Mäntele, W. Infrared reflectometry of skin: Analysis of backscattered light from different skin layers. *Spectrochim. Acta A* **2017**, *184*, 220–227. [CrossRef]
19. Lundsgaard-Nielsen, S.M.; Pors, A.; Banke, S.O.; Henriksen, J.E.; Hepp, D.K.; Weber, A. Critical-depth Raman spectroscopy enables home-use non-invasive glucose monitoring. *PLoS ONE* **2018**, *13*, e0197134. [CrossRef] [PubMed]
20. Heise, H.M.; Marbach, R.; Bittner, A.; Koschinsky, T. Clinical chemistry and near infrared spectroscopy: Multicomponent assay for human plasma and its evaluation for the determination of blood substrates. *J. Near Infrared Spectrosc.* **1998**, *6*, 361–374. [CrossRef]
21. Amerov, A.K.; Chen, J.; Small, G.W.; Arnold, M.A. Scattering and absorption effects in the determination of glucose in whole blood by nears-infrared spectroscopy. *Anal. Chem.* **2005**, *77*, 4567–4594. [CrossRef] [PubMed]
22. Ren, M.; Arnold, M.A. Comparison of multivariate calibration models for glucose, urea, and lactate from near-infrared and Raman spectra. *Anal. Bioanal. Chem.* **2007**, *387*, 879–888. [CrossRef]
23. Li, B.-Y.; Kasemsumran, S.; Hu, Y.; Liang, Y.-Z.; Ozaki, Y. Comparison of performance of partial least squares regression, secured principal component regression, and modified secured principal regression for determination of human serum albumin, γ-globulin, and glucose in buffer solutions and in vivo blood glucose quantification by near-infrared spectroscopy. *Anal. Bioanal. Chem.* **2007**, *387*, 603–611.
24. Smith, J.L. The Pursuit of Noninvasive Glucose: Hunting the Deceitful Turkey, 6th ed.; 2018. Available online: https://www.researchgate.net/publication/327101583_The_Pursuit_of_Noninvasive_Glucose_Hunting_the_Deceitful_Turkey_Sixth_Edition; (accessed on 10 January 2021).
25. Heise, H.M. Medical applications of NIR-spectroscopy. In *Near Infrared Spectroscopy—Theory, Spectral Analysis, Instrumentation, and Applications*; Ozaki, Y., Huck, C., Tsuchikawa, S., Engelsen, S.B., Eds.; Springer Nature: Berlin/Heidelberg, Germany, 2020; Chapter 20.
26. Norris, K.H. Some options for making non-invasive measurements on the human body with near infrared spectroscopy. *J. Near Infrared Spectrosc.* **2012**, *20*, 249–254. [CrossRef]
27. Olivieri, A.C.; Faaber, N.M.; Ferré, J.; Boqué, R.; Kalivas, J.H.; Mark, H. Uncertainty estimation and figures of merit for multivariate calibration. *Pure Appl. Chem.* **2006**, *78*, 633–661. [CrossRef]
28. Heise, H.M.; Winzen, R. Chemometrics in near-infrared spectroscopy. In *Near-Infrared Spectroscopy—Principles, Instruments, Applications*; Siesler, H.W., Ozaki, Y., Kawara, S., Heise, H.M., Eds.; Wiley-VCH: Weinheim, Germany, 2002; Chapter 7.
29. Maruo, K.; Yamada, Y. Near-infrared noninvasive blood glucose prediction without using multivariate analyses: Introduction of imaginary spectra due to scattering change in the skin. *J. Biomed. Opt.* **2015**, *20*, 047003. [CrossRef]
30. Marbach, R.; Heise, H.M. Calibration modeling by partial least-squares and principle component regression and its optimization using an improved leverage correction for prediction testing. *Chemom. Intell. Lab. Syst.* **1990**, *9*, 45–63. [CrossRef]
31. Arnold, M.A.; Small, G.W. Noninvasive glucose sensing. *Anal. Chem.* **2005**, *77*, 5429–5439. [CrossRef]
32. Arnold, M.A.; Liu, L.; Olesberg, J.T. Selectivity assessment of noninvasive glucose measurements based on the analysis of multivariate calibration vectors. *J. Diabetes Sci. Technol.* **2007**, *1*, 454–462. [CrossRef] [PubMed]
33. Heise, H.M.; Lampen, P.; Marbach, R. Near-infrared reflection spectroscopy for non-invasive monitoring of glucose—Established and novel strategies for multivariate calibration. In *Handbook of Optical Sensing of Glucose in Biological Fluids and Tissues*; Tuchin, V.V., Ed.; CRC Press: Boca Raton, FL, USA, 2008; Chapter 5.
34. Marbach, R.; Koschinsky, T.; Gries, F.A.; Heise, H.M. Noninvasive blood glucose assay by near-infrared diffuse reflectance spectroscopy of the human inner lip. *Appl. Spectrosc.* **1993**, *47*, 875–881. [CrossRef]
35. Heise, H.M.; Bittner, A.; Marbach, R. Near-infrared reflectance spectroscopy for non-invasive monitoring of metabolites. *Clin. Chem. Lab. Med.* **2000**, *38*, 137–145. [CrossRef] [PubMed]
36. Heise, H.M.; Bittner, A.; Marbach, R. Clinical chemistry and near infrared spectroscopy: Technology for non-invasive glucose monitoring. *J. Near Infrared Spectrosc.* **1998**, *6*, 349–359. [CrossRef]

37. Marbach, R. On Wiener filtering and the physics behind statistical modeling. *J. Biomed. Opt.* **2002**, *7*, 130–147. [CrossRef] [PubMed]
38. Marbach, R. Methods to Significantly Reduce the Calibration Cost of Multichannel Measurement Instruments. US Patent 6,629,041, 30 September 2003.
39. Marbach, R. A new method for multivariate calibration. *J. Near Infrared Spectrosc.* **2005**, *13*, 241–254. [CrossRef]
40. Barimani, S.; Kleinebudde, P. Evaluation of in–line Raman data for end-point determination of a coating process: Comparison of Science–Based Calibration, PLS-regression and univariate data analysis. *Eur. J. Pharm. Biopharm.* **2017**, *119*, 28–3529. [CrossRef]
41. Marbach, R.; Heise, H.M. Optical diffuse reflectance accessory for measurements of skin tissue by near-infrared spectroscopy. *Appl. Opt.* **1995**, *34*, 610–621. [CrossRef] [PubMed]
42. Bittner, A.; Thomaßen, S.; Heise, H.M. In-vivo measurements of skin tissue by near-infrared diffuse reflectance spectroscopy. *Mikrochim. Acta* **1997**, *14*, 429–432.
43. Jacques, S.L.; Pogue, B.W. Tutorial on diffuse light transport. *J. Biomed. Opt.* **2008**, *13*, 041303. [CrossRef]
44. Troy, T.L.; Thennadiel, S.N. Optical properties of human skin in the near infrared wavelength range of 1000 to 2200 nm. *J. Biomed. Opt.* **2001**, *6*, 167–176. [CrossRef] [PubMed]
45. Salomatina, E.; Jiang, B.; Novak, J.; Yaroslavsky, A.N. Optical properties of normal and cancerous human skin in the visible and near-infrared spectral range. *J. Biomed. Opt.* **2006**, *11*, 064026. [CrossRef] [PubMed]
46. Roggan, A.; Beuthan, J.; Schründer, S.; Müller, G. Diagnostik und Therapie mit dem Laser. *Phys. Blätter* **1999**, *55*, 25–30. [CrossRef]
47. Brown, C.D. Discordance between net analyte signal theory and practical multivariate calibration. *Anal. Chem.* **2004**, *76*, 4364–4373. [CrossRef] [PubMed]
48. Bertie, J.E.; Lan, Z. Infrared intensities of liquids XX: The intensity of the OH stretching band of liquid water revisited, and the best current values of the optical constants of $H_2O(l)$ at 25 °C between 15,000 and 1 cm^{-1}. *Appl. Spectrosc.* **1996**, *50*, 1047–1057. Available online: http://www.ualberta.ca/~{}jbertie/JBDownload.htm (accessed on 10 January 2021). [CrossRef]
49. Mennel, F.J.; Mayer, H.; Bischof, F.; Pfeiffer, E.F. Does fall in tissue glucose precede fall in blood glucose? *Diabetologia* **1996**, *39*, 609–612.
50. Sinha, M.; McKeon, K.M.; Parker, S.; Goergen, L.G.; Zheng, H.; El-Khatib, F.H.; Russel, S.J. A comparison of time delay in three continuous glucose monitors for adolescents and adults. *J. Diabetes Sci. Technol.* **2017**, *11*, 1132–1137. [CrossRef] [PubMed]
51. Cobelli, C.; Schiavon, M.; Man, C.D.; Basu, A.; Basu, R. Interstitial fluid glucose is not just a shifted-in-time but a distorted mirror of blood glucose: Insight from an *in silico* study. *Diabetes Technol. Ther.* **2016**, *18*, 505–511. [CrossRef]

biosensors

Article

Using CNN and HHT to Predict Blood Pressure Level Based on Photoplethysmography and Its Derivatives

Xiaoxiao Sun [1,2], Liang Zhou [1,*], Shendong Chang [3] and Zhaohui Liu [1]

[1] Xi'an Institute of Optics and Precision Mechanics of CAS, Xi'an 710119, China; sunxiaoxiao@opt.cn (X.S.); lzh@opt.ac.cn (Z.L.)

[2] University of Chinese Academy of Sciences, Beijing 100049, China

[3] EAIT (Engineering, Architecture and Information Technology Department), University of Queensland, Brisbane 4072, Australia; shendong.chang@uqconnect.edu.au

* Correspondence: zhouliang@opt.ac.cn; Tel.: +86-187-0686-7958

Abstract: According to the WTO, there were 1.13 billion hypertension patients worldwide in 2015. The WTO encouraged people to check the blood pressure regularly because a large amount of patients do not have any symptoms. However, traditional cuff measurement results are not enough to represent the patient's blood pressure status over a period of time. Therefore, there is an urgent need for portable, easy to operate, continuous measurement, and low-cost blood pressure measuring devices. In this paper, we adopted the convolutional neural network (CNN), based on the Hilbert–Huang Transform (HHT) method, to predict blood pressure (BP) risk level using photoplethysmography (PPG). Considering that the PPG's first and second derivative signals are related to atherosclerosis and vascular elasticity, we created a dataset called PPG+; the images of PPG+ carry information on PPG and its derivatives. We built three classification experiments by collecting 582 data records (the length of each record is 10 s) from the Medical Information Mart for Intensive Care (MIMIC) database: NT (normotension) vs. HT (hypertension), NT vs. PHT (prehypertension), and (NT + PHT) vs. HT; the F1 scores of the PPG + experiments using AlexNet were 98.90%, 85.80%, and 93.54%, respectively. We found that, first, the dataset established by the HHT method performed well in the BP grade prediction experiment. Second, because the Hilbert spectra of the PPG are simple and periodic, AlexNet, which has only 8 layers, got better results. More layers instead increased the cost and difficulty of training.

Keywords: blood pressure; photoplethysmography; derivatives of PPG; convolutional neural network; ensemble empirical mode decomposition

Citation: Sun, X.; Zhou, L.; Chang, S.; Liu, Z. Using CNN and HHT to Predict Blood Pressure Level Based on Photoplethysmography and Its Derivatives. *Biosensors* **2021**, *11*, 120. https://doi.org/10.3390/bios11040120

Received: 19 February 2021
Accepted: 9 April 2021
Published: 13 April 2021

1. Introduction

Patients with chronic hypertension will experience serious consequences if it is left untreated, including a range of cardiovascular diseases affecting the heart [1]. But most patients have no obvious symptoms in the early stages of the disease, so it is important to check BP level regularly.

The traditional method of BP measurement uses a cuff-link-type BP meter. The "white coat effect" refers to patients taking it in a medical setting with even less accurate BP than when they take it at home [2,3]. Therefore, a single measurement datum is not enough to reflect the human condition. Continuous measured BP is more accurate than single measured BP [4]. In view of the shortcomings of clinical invasive continuous BP measurement, which is difficult to perform and may cause serious harm to patients, noninvasive continuous BP measurement is of great significance.

With increases in age and changes in the physical condition of human beings, the elasticity of the blood vessel wall will decrease. When the resistance of blood flowing in blood vessels increases, blood pressure will increase accordingly. This is the pathogenesis of hypertension. Currently, there are three most commonly used noninvasive continuous

BP detection methods. The first is the pulse transit time (PTT) theory method, which refers to the time difference of the diffusion of pulse waves between two arteries in a cardiac cycle. When BP in the blood vessels increases dramatically, the vascular tension and speed of arterial pressure waves will increase, leading to shortened PTT [5]. However, obtaining the time difference requires both electrocardiogram (ECG) and pulse wave signals. It is also difficult to acquire the ECG. The second detection method is based on the morphological theory of PPG [4,6,7]. The most common way to obtain the PPG signal is to use a photoelectric sensor to detect changes in light transmitted or reflected by human skin vessels [8]. PPG represents the change of human blood volume and characterizes the systolic and diastolic processes of the heart, which are linked to BP. PPG's first and second derivative signals are related to atherosclerosis and vascular elasticity, which are factors that influence BP. Luo Zhichang et al. [9] found that the tidal wave of the pulse wave signal of high blood pressure will gradually approach the main wave, finally merge with it, and even exceed the main wave. Therefore, the main wave of the PPG signal with hypertension looks more rounded and curved than that of normotension. Thus, PPG is closely related to BP. Therefore, PPG signal is increasingly applied to predict BP, blood oxygen, respiration rate, and other data [10–12]. The third method combines the characteristic points of the ECG and PPG signals to predict BP [13–15]. The reason for the rise of this method is that, although PPG signals feature points that contain information related to SBP (systolic blood pressure), it is not easy to determine the relationship between PPG and DBP (diastolic blood pressure). Therefore, using PPG alone is bound to lead to inaccurate BP predictions. Accuracy is improved by adding the ECG signal [16]. However, obtaining the ECG signal remains a challenge in research because of current technology limitations.

Among the many studies on PPG signals, research on the derivatives of PPG has attracted our attention. Figure 1 shows PPG and its first and second derivatives. Qawqzeh et al. [17] analyzed the relationship between the first and second derivatives of PPG and atherosclerosis. The second derivative of PPG (SDPTG) was found to be closely related to atherosclerosis and could be used as an assistant technology to detect the disease. There are many factors causing atherosclerosis in the human body, among which hypertension is the most common one. Hypertension and arteriosclerosis cause and affect each other and exist together. Based on this, Mengyang Liu et al. [18] used PPG and its second derivatives to predict noninvasive BP. They retrained the experiment by adding 14 SDPTG features based on the 21 time-scale PPG features, and the experimental results showed that SDPTG can improve the accuracy of BP prediction. However, in real life, the morphological characteristics of PPG and SDPTG vary from person to person, and there are certain difficulties in the calibration and measurement of features, which bring great difficulty to the research. Considering the difficulty mentioned above, many scholars have innovatively used CNN to indirectly identify and extract feature points without manual calibration [19,20], which greatly reduced the time-cost. The self-adaptability of the network structure to extract feature points improves the universality of the experiment and makes the results more convincing. In addition, Liang et al. [21] found that the method using CNN has a higher accuracy than the method using PPG feature point fitting.

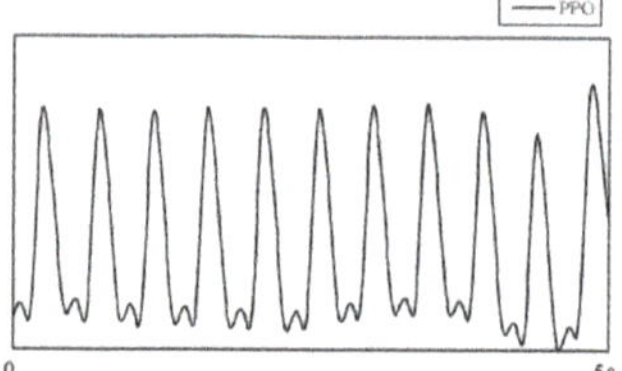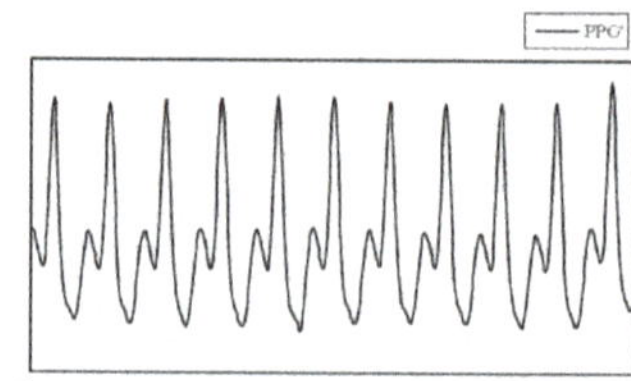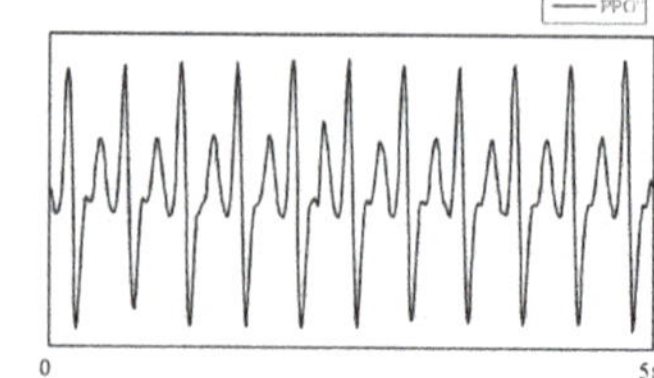

Figure 1. PPG (photoplethysmography) and its first derivative (PPG′) and second derivative (PPG″). The data comes from the MIMIC database.

The purpose of our paper was to use PPG and its first and second derivative signal to predict BP level using a deep learning method. First, we segmented the PPG, extracted the baseline, and regularized the signal. Then the Hilbert–Huang Transform method was used to process the segmented signal and the generated image, and the corresponding BP value of the signal was taken as the label. Finally, the CNN was used to train the dataset. We looked at three questions; the first is to determine whether the training results of the dataset established by the HHT method are good enough on the network. The second is to explore whether more layers in the network are better for our datasets. The third is to know whether the first and second derivatives of PPG carry BP-related information.

2. Materials and Methods

2.1. Dataset Source

Both the PPG and the ABP (arterial blood pressure) data used in this paper come from the MIMIC database. Each signal has 10 s length and a sampling frequency of 125 Hz. In order to improve the generalization performance of the training frame and the generality of the experiment, in addition to the continuous and stable signal, the continuous unstable signal, and the noisy signal were retained. We divided the two signals into 2 parts for 5 s each. After signal processing—including using 0.4–8 Hz Butterworth filter as filtering, normalization, and baseline drift removal—the maximum values of systolic and diastolic blood pressure were extracted from each segment of the ABP signal, and the BP level corresponding to this segment of signal was determined by the 7th Report of the Joint National Committee on Hypertension Prevention, Detection, Evaluation and Treatment (JNC7). By analyzing the values of systolic and diastolic blood pressure (Table 1), JNC7 classifies blood pressure into three categories: normotension (NT), prehypertension (PHT), and hypertension (HT). The BP level shown by the ABP serves as the label for the corresponding segment of the PPG signal.

Table 1. Blood pressure classification according to JNC7.

Classification	Systolic Blood Pressure (mmHg)		Diastolic Blood Pressure (mmHg)
normotension	<120	and	<80
prehypertension	120–139	or	80–89
hypertension	>140	or	>90

2.2. Signal Processing

PPG is a time-domain, nonlinear, and unstable human physiological signal. Therefore, processing the signal can not only retain as much information as possible contained in the original signal but can also provide images that meet the requirements for the deep learning model required by the experiment, which is the main problem we studied in this section.

Slapničar et al. [22] converted PPG and its first and second derivative signals into spectrum diagrams and input three types of spectrum diagrams into the ResNet network for training. This method required a good PPG signal, otherwise the derivative signals would become seriously deformed. Liang et al. [21] used the features of PPG to convert each segment of 5 s signal into a scalogram by using continuous wavelet transform (CWT). CWT is based on the basis function pair, which lacks adaptability and is not easily used to comprehensively describe the complex physiological characteristics of PPG, which is nonlinear and nonstationary. To make up for the shortcomings of previous studies, we decided to use the Hilbert–Huang Transform (HHT), which has the following advantages:

- The nonlinear and nonstationary problems of signals can be solved.
- The motion artifact is effectively removed from the signal.
- Spectra, after transformation, have specific physical meaning related to human physiology.

- One-dimensional signals are converted into two-dimensional signals to facilitate deep learning.

2.3. HHT Based on Ensemble Empirical Mode Decomposition

The Hilbert–Huang Transform was proposed by N. E. Huang [23], who added the empirical mode decomposition (EMD) method on the basis of the Hilbert Transform (HT). EMD (the algorithm flow chart shown in Figure 2) can decompose nonstationary complex signals into intrinsic mode functions (IMFs). Huang et al. believed that any signal is composed of several IMFs; by filtering out the IMF components represented by the high-frequency noise signal and motion artifacts, EMD can realize the smoothing processing of nonstationary signals. By this method, the problem of morphological malformation of the PPG's derivative signals caused by motion artifacts of PPG could be alleviated [24,25]. The IMF components of different frequencies of the decomposed PPG signal represent different physiological meanings. Previous studies have confirmed that different frequency ranges of decomposed IMFs represent different physiological activities. Mitali et al. [26] obtained surrogate respiratory signals at 0.2 Hz to 0.33 Hz. Chuang et al. [27] obtained the related information regarding heart rhythm at 0.4 Hz to 0.9 Hz. IMFs contain the local characteristics of the original signal at different time scales, so they can retain as much of the original information and characteristics of a signal as possible. Since the basis function is obtained by the EMD decomposition of the data itself, compared with the short-time Fourier transform and wavelet decomposition, the EMD method is adaptable, and the signal becomes more direct and intuitive after the Hilbert transform (Figure 3). Hilbert transformation can be performed on the decomposed IMFs to obtain spectra as inputs to the network.

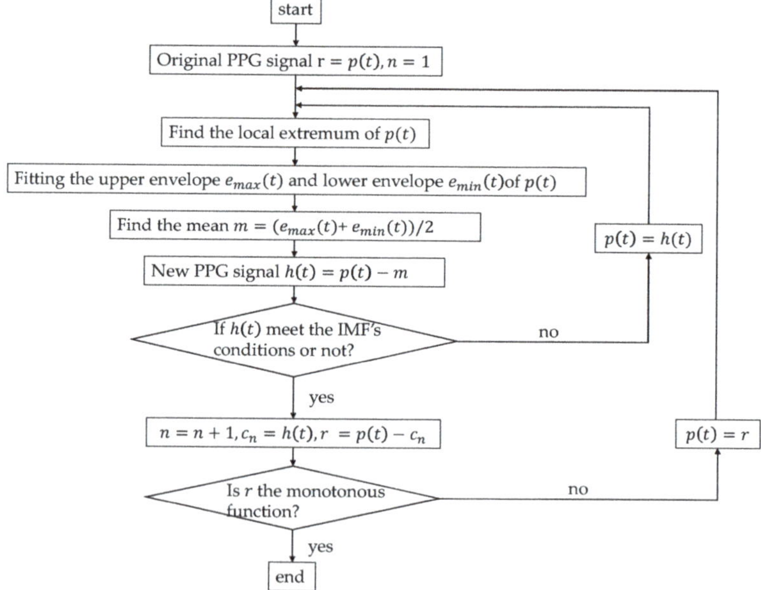

Figure 2. EMD (empirical mode decomposition) processing algorithm. IMF (intrinsic mode function) conditions: (1) In the entire time range, the number of local extreme points and zero points of the function must be the same or differ by one at most. (2) At any point in time, the average value of the envelope of the local maximum of the function (upper envelope) and the envelope of the local minimum (lower envelope) must be zero.

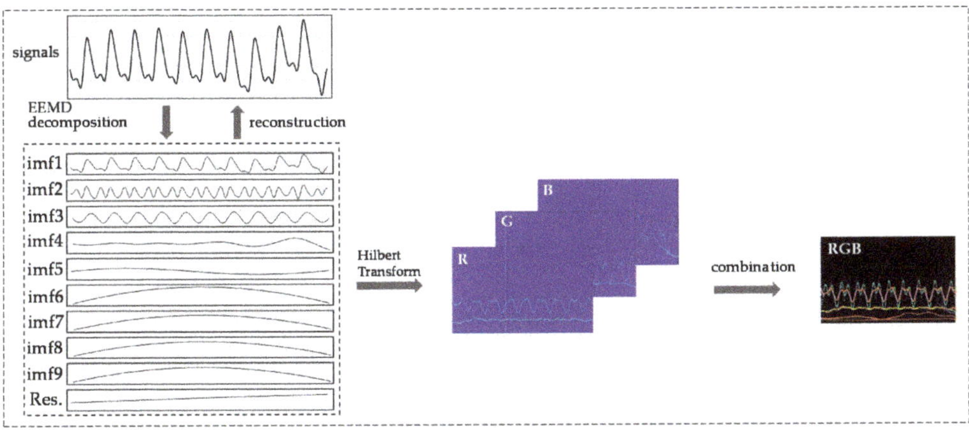

Figure 3. Process of processing PPG signal and its derivatives by the HHT (Hilbert–Huang transform) method.

However, when processing noise signals and intermittent signals, the EMD method will lead to the phenomenon of mode mixing, which seriously affects the accuracy of signal decomposition. To remedy this disadvantage, Huang et al. [28] added white Gaussian noise into the whole time-frequency space for several amounts of time, and several mean values of IMF components were obtained as the final result, which was called EEMD (ensemble empirical mode decomposition). The mode aliasing phenomenon was effectively suppressed by the EEMD method. Besides, this method could be used to remove the motion artifact of PPG signals effectively. To improve the generalization performance of the model, in addition to the serious baseline drift of some PPG signals, we selected some signals with high noise. Therefore, in this paper, we adopt the EEMD method, because this method is better than the EMD method to deal with noisy signals.

The signal processed by the EEMD method is still a one-dimensional vector. Compared with this data format, the features of the signal shown by the two-dimensional image are more prominent. Combining the advantages of convolutional neural networks in the field of image recognition and its own characteristics, images with more prominent feature points require fewer layers, fewer calculations, and less complexity. Furthermore, the spectra of the PPG signal after HHT are related to the time-instantaneous frequency–energy distribution map. Observing the spectra of normotension and hypertension samples (Figure 4b), we found that the instantaneous frequency of the energy distribution in the spectra of the latter is higher than that of the former. Such significant information cannot be obtained using a one-dimensional vector to represent the PPG signal. Therefore, we used a two-dimensional signal as the input of the convolutional neural network.

We built the PPG+ dataset [29]; the PPG's Hilbert spectrogram is the red channel of the RGB image. The spectrograms of the PPG's first and second derivatives are the other two channels (Figure 3) to improve the information dimension and enhance the image's extractable features. These three channels combine to form an RGB image with resolution of 1247 × 770 (Figure 4c); the image length is 5 s (0–5 s), and the width is 13 Hz (0–13 Hz). The sampling quality of the images is 24 bytes, which is 8 bits for each channel and 256 levels. As the input to the network, HT spectra require preprocessing such as random aspect ratio cropping, random flipping, and center cropping before entering the network. This approach is important to enhance the data. Then, we normalized the processed image, and the image size became 3 × 224 × 224. Furthermore, the control experiment's inputs were only the PPG signals' spectrograms.

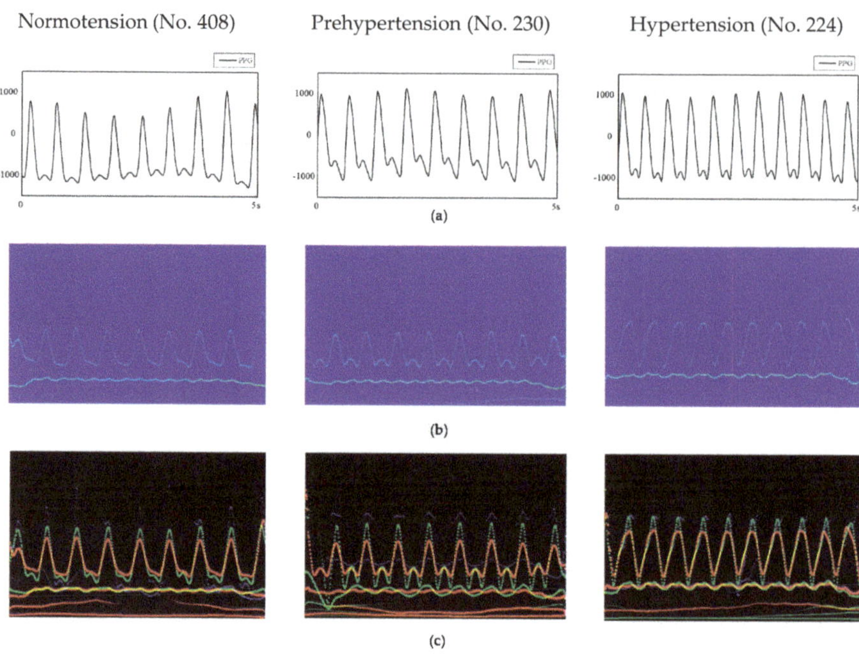

Figure 4. HHT spectrograms of three patients in different time periods. (**a**) PPG signals. (**b**) HHT spectra of PPG signals. (**c**) RGB images of PPG and its first and second derivative (PPG+) spectrogram combination. No. 408, No. 230 and No. 224 are the numbers of the patients.

2.4. Model Fine-Tuning

Transfer learning is a branch of machine learning. The aim is to deal with a new problem by transferring it to a problem that has been resolved. Transfer learning can greatly reduce the experimental costs and training time of deep learning models and also can be applied to the problem of small datasets. Many studies have shown that transfer learning can improve the generalization ability of models [30].

Fine-tuning is a widespread practice in transfer learning, and it is suitable for small datasets. One common practice is to remove the last layer of the pretrained network and replace it with a new softmax layer, which relates to a specific problem. Considering the characteristics of the model's inputs, we used the AlexNet [31], ResNet18 [32], ResNet34 and GoogLeNet to train the data. AlexNet has five layers of convolutional layers and three layers of fully connected layers. The ReLU, as the activation function, solves the gradient dispersion problem. ResNet adds a residual block through a short-circuit mechanism between every two layers of the ordinary network, transforms the difficult identity mapping problem into a residual problem, solves network performance degradation with increasing depth, and effectively alleviates the gradient dispersion or gradient explosion phenomenon of deep neural networks. GoogLeNet has 22 layers. It connects multiple inception blocks in series with other layers to improve the expressive ability of the network. Compared with other networks, GoogLeNet has fewer training parameters and faster convergence speed. The input size of GoogLeNet is 3*299*299, which is different than the other three networks.

In our work, the fine-tuning method removed the last fully connected layer when training with AlexNet, added Dropout to ignore some neurons to prevent the model from overfitting randomly, and finally added a fully connected layer with two outputs. When it came to ResNet18 and ResNet34, the specific method was to reinitialize the last fully connected layer and a linear layer with 512 input features and two output features. The parameters did not need to be trained from scratch but only fine-tuned based on the original model parameters, which dramatically improved the training efficiency and saved on training costs. Figure 5 shows the preprocessing of the signal and the classification experiment.

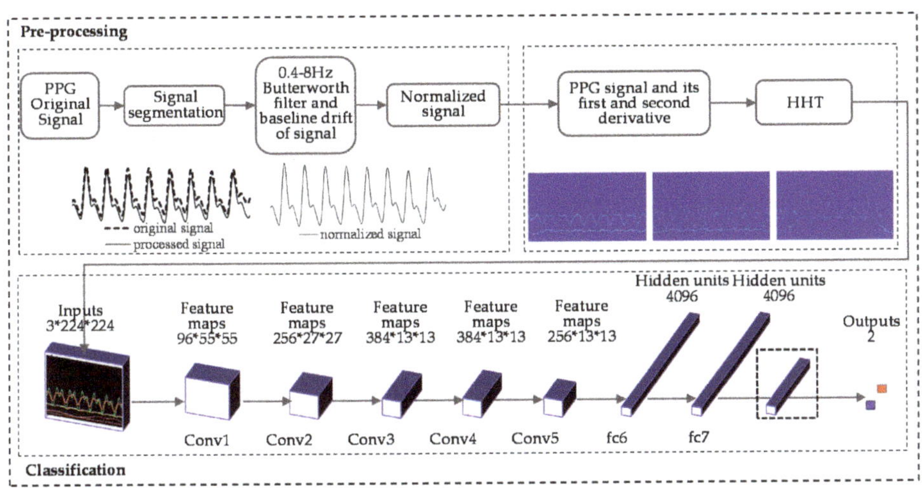

Figure 5. PPG signals processing procedure and the classification experiment (AlexNet).

2.5. Classification Experiment

We built three classification experiments by collecting 582 data records from the MIMIC database: HT (hypertension) vs. NT (normotension), HT vs. PHT (prehypertension), and (HT + PHT) vs. NT. Two datasets, PPG and PPG+, were trained on the following networks with different layers: AlexNet, ResNet18, GoogLeNet, and ResNet34. Three two-classification experiments were performed on each dataset for every network. We designed the experiment for three purposes: First, to explore whether using our dataset to predict blood pressure levels could get better results. Second, to verify whether the PPG+ dataset containing the information of the PPG itself and its first and second derivatives was better than the PPG dataset for blood pressure prediction. Third, to consider whether the higher the number of layers, the better the prediction results would be.

The programming language used for the experiment was Python; the library used to call the model structure was PyTorch, and the model was trained on Anaconda. Code was run using a desktop with an Intel i5-8500 as the CPU, 8GB RAM, and AMD Raden R5 430 as the graphic card. After testing and optimizing the model many times, we chose Adam as the optimizer of the model, and the learning rate parameter of the optimizer was set to 0.001. Since our dataset was relatively small, we divided all data into training sets and test sets in to the ratio 7:3. To improve the generalization performance of the model and get more accurate results, we applied k-fold cross-validation to train the dataset and averaged the results. Taking into account the limitations of computer performance and the time-cost required for cross-validation, the value of k here was 5. We performed 25 epochs for each fold. The fine-tuned AlexNet used a total of 57.01M parameters, and the amount of calculation (floating point operations) was 0.71 GMAC (Memory Access Cost). After testing, the model worked best when the dropout value was set to 0.6 (for AlexNet).

3. Results

In this paper, we randomly selected 582 data records of ABP and PPG from the MIMIC database; each record was 10 s in length. We divided the 2 signals into 2 segments of 5 s each. Then the signal was filtered with a 0.4–8 Hz Butterworth filter and normalized.

The EEMD method was used to remove the signal's motion artifacts and baseline drift. According to the blood pressure classification standard from JNC7, the ABP of each segment was analyzed and then classified as the label of the corresponding PPG signal. Then we performed Hilbert processing on the PPG and made two datasets of PPG and PPG+. The images of the PPG+ dataset contain the first and second derivatives of PPG in addition to the signal itself. Finally, we put these two datasets into four different layers of network models for training. The results (Table 2) are represented by the F1 score, TPR (true positive rate), and TNR (true negative rate).

Table 2. Classification performance of the proposed deep learning method. TPR stands for true positive rate and refers to the sensitivity of the model. TNR is short for true negative rate and refers to the specificity of the model.

CNN	Layers	Trail	PPG			PPG+		
			F1 Score	TPR	TNR	F1 Score	TPR	TNR
AlexNet	8	NT vs. HT	96.33%	94.69%	97.93%	98.90%	99.27%	98.31%
		NT vs. PHT	80.35%	78.98%	84.08%	85.80%	95.26%	71.88%
		(NT + PHT) vs. HT	90.79%	90.39%	91.36%	93.54%	95.32%	91.74%
ResNet18	18	NT vs. HT	93.94%	94.51%	93.17%	94.09%	95.36%	92.33%
		NT vs. PHT	82.34%	82.85%	82.89%	84.37%	84.51%	84.59%
		(NT + PHT) vs. HT	87.35%	90.06%	84.62%	88.52%	88.87%	88.92%
GoogLeNet	22	NT vs. HT	89.48%	88.79%	90.61%	89.24%	90.19%	88.26%
		NT vs. PHT	78.05%	77.73%	79.95%	80.03%	80.79%	77.47%
		(NT + PHT) vs. HT	84.04%	81.51%	88.30%	83.46%	83.76%	83.14%
ResNet34	34	NT vs. HT	93.04%	93.04%	93.41%	94.01%	93.85%	94.26%
		NT vs. PHT	81.33%	81.75%	82.75%	84.77%	83.71%	86.34%
		(NT + PHT) vs. HT	86.76%	88.00%	83.23%	88.39%	87.15%	90.19%

We reached the following conclusions from the table:

- Our dataset performed well on convolutional neural networks, especially AlexNet. In the NT vs. HT experiment, the F1 score was close to 99%, followed by (NT + PHT) vs. HT, and the training effect was the worst in the NT vs. PHT. The two datasets showed similar results on all the mentioned CNNs.
- When it came to the same network on the different datasets, the CNN performed better on the PPG+ dataset than on the PPG dataset. The improvement of the F1 score was evident in the NT vs. HT experiment of AlexNet.
- Compared to the results of different CNNs on the same experiment on the PPG+ dataset, we found that, although AlexNet has the least number of layers, the F1 score was the highest on the experiment of NT vs. HT and (NT + PHT) vs. HT. As the number of convolutional layers increased, the result did not improve but did decrease. For NT vs. HT, the effect of ResNet34 of 34 layers was improved compared with ResNet18 of 18 layers, but only a little.

From the results, the F1 score of AlexNet for our dataset was high: 98.90%. In the NT vs. HT experiment of AlexNet by using the PPG+ dataset as an example, TPR refers to the sensitivity of the model, which means that the number of normal blood pressure samples predicted by the model accounting for 99.27% of the actual normal blood pressure samples. TNR refers to the specificity of the model, which presents the sample size of hypertension predicted by the model to be 98.31% of the actual sample size of hypertension.

The ROC (receiver operating characteristic) curve (Figure 6) characterizes the generalization ability of the learner from the perspective of threshold selection. The closer it is to the upper left corner, the lower the classification error rate of the model, the better the threshold at this time, and the fewer false positives and false negatives there are. We found that the ROC curve of the experiment of NT vs. HT is closest to the upper left corner, which means our network performed best on it.

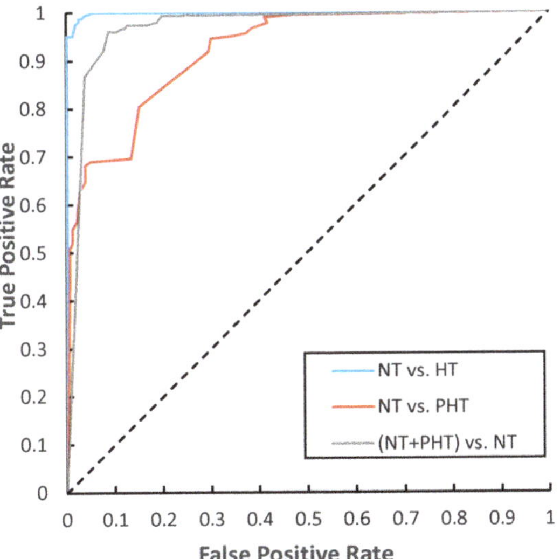

Figure 6. The ROC (receiver operating characteristic) curve of the three classification trials of AlexNet.

Comparing the classification accuracy of AlexNet on the training set and the test set (Figure 7), we found the epoch_accuracy curve of the training set and the test set were very close, which means that the model performed well on our dataset, and there was neither overfitting nor underfitting. The accuracy curves of NT vs. HT and (NT + PHT) vs. HT were close to stable, indicating that the two have converged and the number of training epochs was sufficient. The accuracy of NT vs. PHT on the test set fluctuates greatly, and it has not converged yet, indicating that 25 epochs is not enough for this experiment. However, through the performance on the training set, we can basically infer the future trend of the accuracy curve of the test set. Therefore, upgrading hardware equipment and increasing the number of epochs in the experiment to improve the training effect is part of our future work.

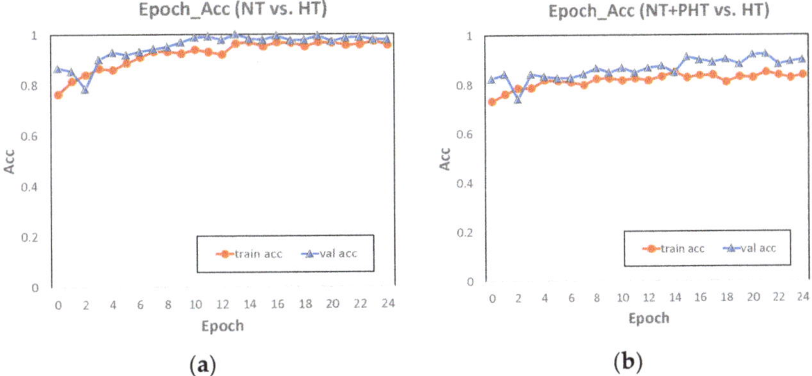

Figure 7. *Cont.*

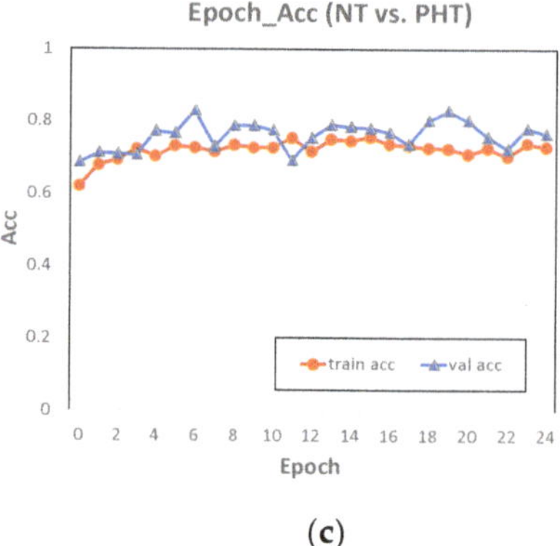

(**c**)

Figure 7. The diagram of epoch and accuracy (referred to as acc in the figure) of the NT vs. HT (**a**), NT + PHT vs. HT (**b**) and NT vs. PHT (**c**) in AlexNet.

4. Discussion

We used a convolutional neural network to predict blood pressure levels using PPG and its first and second derivative signals. The HHT method was applied to remove the baseline and motion artifact of PPG signals, and then we converted the one-dimensional signal into spectra as the input of the network. The blood pressure level represented by ABP was used as the label of each spectrum. According to the information contained in spectra, we established two datasets. One was called PPG+, which contained PPG itself and its first and second derivative signals. The other one was called PPG, which only contained the information in the PPG signal itself. Three binary classification experiments for each dataset were trained on four network models with different layers, AlexNet, ResNet18, GoogLeNet, and ResNet34. This study found that, although AlexNet had a small number of layers, the training results were still the best on both datasets. The fine-tuned AlexNet performed well in the NT vs. HT experiment, with an F1 score of 98.90%. The result of experiments confirmed that the first and second derivatives of PPG could improve training accuracy.

We noticed that the model performed best in NT vs. HT and performed poorly in NT vs. PHT. ROC showed that NT vs. HT was closest to the upper left corner; the distributions of these two types of data were far apart and easier to separate. The model performed poorly in the HT vs. PHT experiment because the two types of data distribution were much closer. In comparison, we also found that the PPG+ dataset had better performance than the PPG dataset in the BP classification model, especially in the NT vs. HT experiment. This confirmed that the first and second derivatives of the PPG signals carry information about BP.

The first derivative of the PPG signal represented the blood flow velocity in the aorta, and the second derivative signal represented the change in blood flow velocity (Figure 4c), which is decided by the blood viscosity and elasticity of the blood vessel wall. The hypertensive patients' blood pressure is high, so when the aortic valve opens, blood flows into the aorta quicker. If the blood vessel elasticity is lacking, the PPG signal's descending branch will be steeper than in ordinary people. The second derivative of the PPG signal happens to reflect this. This is the reason that adding PPG derivative information to the dataset can improve the accuracy of BP prediction.

In order to improve on the weak points of previous studies (Table 3), this paper adopted the EEMD method to process signals. The resulting HHT spectra contained the physical meaning of a particular PPG signal. The PPG signal represents the change of blood volume in human blood vessels, and the derivatives of PPG signal contain information related to blood flow. Therefore, we added the first and second derivative signals of PPG signals to the model to train on simultaneously.

Table 3. Comparison with well-established related work in terms of data source, feature, signal processing, and method.

Author	Data Source	Feature	Signal Process	Method
Slapničar et al. [22]	MIMIC (510 subjects over 700 h)	PPG, PPG′, PPG″	Spectro-temporal	ResNet
Liang et al. [21]	MIMIC (121 data records for 120 s each)	PPG	CWT	GoogLeNet
Our work	MIMIC (582 data records for 10 s each)	PPG, PPG′, PPG″	EEMD	AlexNet

The purpose of converting a one-dimensional signal into a spectrum was not only to satisfy the input format of the network model but also to make the information contained in the PPG signal more prominent. The experimental results showed that the smaller the number of layers, the better the training effect on our dataset. This result proved our approach: the feature points become more obvious after the PPG one-dimensional signal is converted into spectra, and better results can be learned without too many layers of convolutional neural networks. This can reduce the cost of model training. Compared with image recognition and classification problems such as cats and dogs, the input images in this paper had fewer feature points and lower complexity. This might be one of the reasons that the accuracy decreased or remained unchanged with more network layers in the experimental results.

Since the dataset mentioned in the paper [21,22] was not available, we used GoogLeNet to train the dataset for comparison. The F1 score of NT vs. HT in the PPG+ dataset was 89.24%. Although the result was poor compared with previous research, this result could not be evaluated arbitrarily because of the different datasets. What was certain, however, was that this result again confirmed our point that our dataset performed better on a network with fewer layers. AlexNet with only eight layers performed better on both datasets than some other networks with more layers. On the contrary, GoogLeNet, which has many layers and many complex inception blocks, did not perform well on our datasets.

5. Conclusions

In this paper, we used the HHT method to establish a new dataset. The fine-tuned AlexNet performed well on our dataset. The F1 score of the NT vs. HT binary classification experiment can reach 98.90%. The signals processed by HHT have specific physical meanings and obvious feature points, which were conducive to the learning of neural networks. By comparing the performance of blood pressure classification experiments in different network models, our study proved that the derivatives of PPG carry important information on blood pressure, which means that PPG and its derivatives can be used to replace the combination of ECG and PPG for blood pressure prediction. We also found that a network structure with fewer layers had a better performance on our dataset. This can reduce the amount of calculation and the time-cost of network training. We combined the EEMD method with deep learning, providing new ideas for modern medical health testing while providing a noninvasive, fast, and low-cost BP level assessment method for families and low- and middle-income countries. However, this technology still has lots of room for improvement. Our next target will focus on how to improve the classification accuracy and how to predict BP values through deep learning and will explore more information related to physiological activities from PPG and its derivative signals.

Author Contributions: X.S. did validation, formal analysis, project investigations, data curation, writing–original draft preparation, graphic visualization and code writing. X.S. and S.C. reviewed and edited the manuscript together. X.S., S.C. and L.Z. collected resources. Z.L. and L.Z. are the project administrators. All authors have read and agreed to the published version of the manuscript.

Funding: This research was funded by the National Natural Science Foundation of China, grant number 61805275.

Institutional Review Board Statement: Ethical review and approval were waived for this study, due to all data are collected from open-downloadable datasets.

Informed Consent Statement: Patient consent was waived due to all data are collected from open-downloadable datasets which written informed consent should be already obtained by sharing institution and all participant agreed to disclose their data to research.

Data Availability Statement: The signal data in the MIMIC database were downloaded from the following website: https://archive.physionet.org/cgi-bin/atm/ATM. The dataset we created is available at the following link: http://figshare.com/articles/figure/Blood_pressure_classification_experiment/13498422/1.

Conflicts of Interest: The authors declare no competing financial or nonfinancial interest.

References

1. Mills, K.T.; Stefanescu, A.; He, J. The Global Epidemiology of Hypertension. *Nat. Rev. Nephrol.* **2020**, *16*, 223–237. [CrossRef] [PubMed]
2. Wang, X.; Shuai, W.; Peng, Q.; Li, J.; Li, P.; Cheng, X.; Su, H. White Coat Effect in Hypertensive Patients: The Role of Hospital Environment or Physician Presence. *J. Am. Soc. Hypertens.* **2017**, *11*, 498–502. [CrossRef]
3. Feitosa, A.D.M.; Mota-Gomes, M.A.; Barroso, W.S.; Miranda, R.D.; Barbosa, E.C.D.; Pedrosa, R.P.; Oliveira, P.C.; Feitosa, C.L.D.M.; Brandão, A.A.; Lima-Filho, J.L.; et al. Blood Pressure Cutoffs for White-Coat and Masked Effects in a Large Population Undergoing Home Blood Pressure Monitoring. *Hypertens. Res.* **2019**, *42*, 1816–1823. [CrossRef]
4. Yang, S.; Zhang, Y.; Morgan, S.; Cho, D.S.-Y.; Correia, R.; Wen, L. Cuff-Less Blood Pressure Measurement Using Fingertip Photoplethysmogram Signals and Physiological Characteristics. In Proceedings of the Optics in Health Care and Biomedical Optics VIII, Beijing, China, 23 October 2018; Luo, Q., Li, X., Tang, Y., Gu, Y., Eds.; SPIE: Ningbo, China, 2018; p. 116.
5. Smith, R.P.; Argod, J.; Pepin, J.-L.; Levy, P.A. Pulse Transit Time: An Appraisal of Potential Clinical Applications. *Thorax* **1999**, *54*, 452–457. [CrossRef] [PubMed]
6. Teng, X.F.; Zhang, Y.T. Continuous and Noninvasive Estimation of Arterial Blood Pressure Using a Photoplethysmographic Approach. In Proceedings of the 25th Annual International Conference of the IEEE Engineering in Medicine and Biology Society, (IEEE Cat. No.03CH37439), Cancun, Mexico, 17–21 September 2003; Volume 4, pp. 3153–3156.
7. Goudarzi, R.H.; Somayyeh Mousavi, S.; Charmi, M. Using Imaging Photoplethysmography (IPPG) Signal for Blood Pressure Estimation. In Proceedings of the 2020 International Conference on Machine Vision and Image Processing (MVIP), Qom, Iran, 18–20 February 2020; pp. 1–6.
8. Elgendi, M.; Fletcher, R.; Liang, Y.; Howard, N.; Lovell, N.H.; Abbott, D.; Lim, K.; Ward, R. The Use of Photoplethysmography for Assessing Hypertension. *NPJ Digit. Med.* **2019**, *2*, 60. [CrossRef] [PubMed]
9. Luo, Z. *Engineering Analysis and Clinical Application of Pulse Wave*; Science Press: Beijing, China, 2006; ISBN 978-7-03-017048-4.
10. Reiss, A.; Schmidt, P.; Indlekofer, I.; Van Laerhoven, K. PPG-Based Heart Rate Estimation with Time-Frequency Spectra: A Deep Learning Approach. In Proceedings of the 2018 ACM International Joint Conference and 2018 International Symposium on Pervasive and Ubiquitous Computing and Wearable Computers—UbiComp '18, Singapore, 8–12 October 2018; ACM Press: Singapore, 2018; pp. 1283–1292.
11. Pereira, T.; Tran, N.; Gadhoumi, K.; Pelter, M.M.; Do, D.H.; Lee, R.J.; Colorado, R.; Meisel, K.; Hu, X. Photoplethysmography Based Atrial Fibrillation Detection: A Review. *NPJ Digit. Med.* **2020**, *3*, 1–12. [CrossRef] [PubMed]
12. Haneishi, H.; Nishidate, I.; Nakano, K.; Niizeki, K.; Aizu, Y.; McDuff, D.J. Evaluation of Arterial Oxygen Saturation Using RGB Camera-Based Remote Photoplethysmography. In Proceedings of the Optical Diagnostics and Sensing XVIII: Toward Point-of-Care Diagnostics, San Francisco, CA, USA, 27 January–1 February 2018; Coté, G.L., Ed.; SPIE: San Francisco, CA, USA; p. 35.
13. Zhang, Q.; Zhou, D.; Zeng, X. Highly Wearable Cuff-Less Blood Pressure and Heart Rate Monitoring with Single-Arm Electrocardiogram and Photoplethysmogram Signals. *Biomed. Eng. Online* **2017**, *16*, 23. [CrossRef] [PubMed]
14. Senturk, U.; Yucedag, I.; Polat, K. Repetitive Neural Network (RNN) Based Blood Pressure Estimation Using PPG and ECG Signals. In Proceedings of the 2018 2nd International Symposium on Multidisciplinary Studies and Innovative Technologies (ISMSIT), Ankara, Turkey, 19–21 October 2018; pp. 1–4.

15. Singla, M.; Sistla, P.; Azeemuddin, S. Cuff-Less Blood Pressure Measurement Using Supplementary ECG and PPG Features Extracted Through Wavelet Transformation. In Proceedings of the 2019 41st Annual International Conference of the IEEE Engineering in Medicine and Biology Society (EMBC), Berlin, Germany, 23–27 July 2019; pp. 4628–4631.
16. Liang, Y.; Chen, Z.; Ward, R.; Elgendi, M. Hypertension Assessment via ECG and PPG Signals: An Evaluation Using MIMIC Database. *Diagnostics* **2018**, *8*, 65. [CrossRef] [PubMed]
17. Qawqzeh, Y. Assessment of Atherosclerosis in Erectile Dysfunction Subjects Using Second Derivative of Photoplethysmogram. *Sci. Res. Essays* **2012**, *7*. [CrossRef]
18. Liu, M.; Po, L.-M.; Fu, H. Cuffless Blood Pressure Estimation Based on Photoplethysmography Signal and Its Second Derivative. *IJCTE* **2017**, *9*, 202–206. [CrossRef]
19. Yan, C.; Li, Z.; Zhao, W.; Hu, J.; Jia, D.; Wang, H.; You, T. Novel Deep Convolutional Neural Network for Cuff-Less Blood Pressure Measurement Using ECG and PPG Signals. Available online: https://pubmed.ncbi.nlm.nih.gov/31946273/ (accessed on 18 January 2021).
20. Mansouri, S.; Lowe, A.; Hosseini, H.; Baig, M. Blood Pressure Estimation from Electrocardiogram and Photoplethysmography Signals Using Continuous Wavelet Transform and Convolutional Neural Network. In Proceedings of the CONF-IRM 2019 Proceedings, Auckland, New Zealand, 27–29 May 2019.
21. Liang, Y.; Chen, Z.; Ward, R.; Elgendi, M. Photoplethysmography and Deep Learning: Enhancing Hypertension Risk Stratification. *Biosensors* **2018**, *8*, 101. [CrossRef] [PubMed]
22. Slapničar, G.; Mlakar, N.; Luštrek, M. Blood Pressure Estimation from Photoplethysmogram Using a Spectro-Temporal Deep Neural Network. *Sensors* **2019**, *19*, 3420. [CrossRef] [PubMed]
23. Huang, N.E.; Shen, Z.; Long, S.R.; Wu, M.C.; Shih, H.H.; Zheng, Q.; Yen, N.-C.; Tung, C.C.; Liu, H.H. The Empirical Mode Decomposition and the Hilbert Spectrum for Nonlinear and Non-Stationary Time Series Analysis. *Proc. R. Soc. Lond. A* **1998**, *454*, 903–995. [CrossRef]
24. Pang, B.; Liu, M.; Zhang, X.; Li, P.; Chen, H. A Novel Approach Framework Based on Statistics for Reconstruction and Heartrate Estimation from PPG with Heavy Motion Artifacts. *Sci. China Inf. Sci.* **2018**, *61*, 022312. [CrossRef]
25. Raghuram, M.; Madhav, K.V.; Krishna, E.H.; Komalla, N.R.; Sivani, K.; Reddy, K.A. HHT Based Signal Decomposition for Reduction of Motion Artifacts in Photoplethysmographic Signals. In Proceedings of the 2012 IEEE International Instrumentation and Measurement Technology Conference Proceedings, Graz, Austria, 13–16 May 2012; pp. 1730–1734.
26. Ambekar, M.R.; Prabhu, S. A Novel Algorithm to Obtain Respiratory Rate from the PPG Signal. *Int. J. Comput. Appl.* **2015**, *126*, 9–12. [CrossRef]
27. Chuang, S.-Y.; Liao, J.-J.; Chou, C.-C.; Chang, C.-C.; Fang, W.-C. Spectral Analysis of Photoplethysmography Based on EEMD Method. In Proceedings of the 2015 IEEE International Conference on Consumer Electronics, Taipei, Taiwan, 6–8 June 2015; pp. 224–225.
28. Wu, Z.; Huang, N.E. Ensemble Empirical Mode Decomposition: A Noise-Assisted Data Analysis Method. *Adv. Adapt. Data Anal.* **2009**, *1*, 1–41. [CrossRef]
29. Sun, X. Blood Pressure Classification Experiment 2020. Available online: http://figshare.com/articles/figure/Blood_pressure_classification_experiment/13498422/1 (accessed on 19 March 2021).
30. Fuzhen, Z.; Ping, L.; Qing, H. Survey on Transfer Learning Research. *J. Softw.* **2015**, *26*, 26–39.
31. Krizhevsky, A.; Sutskever, I.; Hinton, G.E. ImageNet Classification with Deep Convolutional Neural Networks. *Commun. ACM* **2017**, *60*, 84–90. [CrossRef]
32. He, K.; Zhang, X.; Ren, S.; Sun, J. Deep Residual Learning for Image Recognition. In Proceedings of the 2016 IEEE Conference on Computer Vision and Pattern Recognition (CVPR), Las Vegas, NV, USA, 27–30 June 2016; pp. 770–778.

biosensors

Article

Novel Micro-Nano Optoelectronic Biosensor for Label-Free Real-Time Biofilm Monitoring

Giuseppe Brunetti [1], Donato Conteduca [1,2], Mario Nicola Armenise [1] and Caterina Ciminelli [1,*

[1] Optoelectronics Laboratory, Department of Electrical and Information Engineering, Polytechnic University of Bari, 70125 Bari, Italy; giuseppe.brunetti@poliba.it (G.B.); donato.conteduca@york.ac.uk (D.C.); marionicola.armenise@poliba.it (M.N.A.)

[2] Photonics Group, Department of Physics, University of York, Heslington, York YO10 5DD, UK

* Correspondence: caterina.ciminelli@poliba.it

Abstract: According to the World Health Organization forecasts, AntiMicrobial Resistance (AMR) is expected to become one of the leading causes of death worldwide in the following decades. The rising danger of AMR is caused by the overuse of antibiotics, which are becoming ineffective against many pathogens, particularly in the presence of bacterial biofilms. In this context, non-destructive label-free techniques for the real-time study of the biofilm generation and maturation, together with the analysis of the efficiency of antibiotics, are in high demand. Here, we propose the design of a novel optoelectronic device based on a dual array of interdigitated micro- and nanoelectrodes in parallel, aiming at monitoring the bacterial biofilm evolution by using optical and electrical measurements. The optical response given by the nanostructure, based on the Guided Mode Resonance effect with a Q-factor of about 400 and normalized resonance amplitude of about 0.8, allows high spatial resolution for the analysis of the interaction between planktonic bacteria distributed in small colonies and their role in the biofilm generation, calculating a resonance wavelength shift variation of 0.9 nm in the presence of bacteria on the surface, while the electrical response with both micro- and nanoelectrodes is necessary for the study of the metabolic state of the bacteria to reveal the efficacy of antibiotics for the destruction of the biofilm, measuring a current change of 330 nA when a 15 µm thick biofilm is destroyed with respect to the absence of biofilm.

Keywords: bacteria biofilm; optoelectronic device; antimicrobial resistance; biosensing

Citation: Brunetti, G.; Conteduca, D.; Armenise, M.N.; Ciminelli, C. Novel Micro-Nano Optoelectronic Biosensor for Label-Free Real-Time Biofilm Monitoring. *Biosensors* **2021**, *11*, 361. https://doi.org/10.3390/bios11100361

Received: 26 August 2021
Accepted: 27 September 2021
Published: 29 September 2021

Publisher's Note: MDPI stays neutral with regard to jurisdictional claims in published maps and institutional affiliations.

1. Introduction

Bacterial infections represent one of the leading causes of death in developing nations [1]. The infections are caused by food poisoning, which is often related to water contamination or improper food preparation [2]. Furthermore, the large overuse and/or misuse of antibiotics is causing a rapid growth in AntiMicrobial Resistance (AMR) worldwide [3]. AMR is developed when the bacteria adapt to and resist antibiotics treatments, which become ineffective to counteract a bacterial infection that can grow and spread in a large community through direct contact, food, or the environment [4,5]. As a result of the lack of powerful antibiotics, many bacterial infections, such as pneumonia, tuberculosis, and gonorrhoea, are becoming more difficult to eradicate with a consequent higher mortality rate [6]. According to [7], the cost of AMR on public health is up to 100 trillion USD, and AMR is expected to become the leading cause of death worldwide, with over 10 million annually predicted by 2050. These consequences highlight that AMR is a widespread social problem that cannot be underestimated or neglected anymore due to the large and rising number of people potentially affected.

Many bacterial infections are caused by the non-eradication of bacterial biofilm, which can be several times more resistant to antibiotics compared to planktonic bacteria [8,9]. This behavior is strictly correlated to the intrinsic nature of the biofilm, which consists of densely packed microbial cells that can grow and surround themselves with a self-produced Extracellular Matrix (*ECM*). The *ECM* is composed of proteins, polysaccharides, and nucleic acids that protect the bacterial biofilm from the environment, so making it very resistant to external agents, such as antibiotics [10]. Moreover, a biofilm may include different bacteria, and this makes the dissolution of biofilms more challenging [11]. In fact, it has been demonstrated that even if a biofilm is treated by an antibiotic that is efficient for a specific planktonic bacterium or small communities of bacteria with a concentration much higher than Minimum Biofilm Inhibitory Concentration (*MIC*), which represents the lowest concentration of drug to prevent the bacteria growth, the biofilm structure can be completely unaltered, showing a continuous growth process also after the treatment [12].

2. Techniques for the Bacteria Detection and Analysis

To date, the most widely used diagnostics to detect the presence of bacteria and analyze their evolution under several antibiotics' treatment is the plate-count method, which is based on the growth of bacteria on an agar plate [13,14]. However, this technique is time consuming (24–72 h), because it requires many cell-division cycles as well as expert users for the sample preparation and final analysis. The large time delay is the most significant bottleneck of such a technique because several infections, such as sepsis, require an immediate measure, also to avoid the formation of a biofilm [15]. The antibiotics are commonly administered in the clinicians' experience with a not-negligible failure rate, so possibly leading to an outbreak of the resistance rather than by carrying out an accurate diagnosis. Thus, novel diagnostic techniques that can rapidly detect and identify bacteria and confirm the presence of a biofilm, ideally within 30 min, are needed. Furthermore, these techniques should also enable study of the efficiency of antibiotics with a real-time analysis during the treatment in order to define the most powerful antibiotic, the best concentration, and the administration time for each infection [16].

During the last few years, several approaches have been investigated, mainly with the use of optical techniques, such as Raman Spectroscopy or fluorescence, due to their high resolution and real-time detection of individual bacteria [17–19]. However, these methods are not label-free, becoming inefficient in the presence of bacteria mutations, and allow investigating only a small area. Integrated optical devices have also been used for single bacteria analysis with label-free techniques [20]. In particular, resonant cavities are able to trap and identify single bacteria through the changes of the resonance response [21,22]. Emerging studies with a label-free optical-based approach have demonstrated real-time monitoring of cell attachment and the development of bacteria on the sensor surface [23]. Optical devices have been used to investigate the antimicrobial susceptibility with in vitro studies by adopting antibiotics with concentrations compliant with standard health protocols [23]. However, the simultaneous detection of several bacteria in the whole biofilm volume is still challenging because of the mismatch between the biofilm thickness and evanescent field penetration depth. In fact, the evanescent field of resonant cavities typically extends for few hundreds of nanometers, thus making impossible the detection of multiple bacteria organized in a three-dimensional configuration as a biofilm that can reach a thickness of several microns in the presence of macro-colonies in the maturation phase.

The mechanical trapping of bacteria has also been obtained by using microfluidic devices [24,25], allowing the bacteria localization in specific areas to accurately detect and analyze them with an atomic force microscope (*AFM*) [26]. The *AFM* technique guarantees a very high resolution, also providing relevant information about the metabolic state of bacteria and if they are live/dead, for example by observing a different motility with a change of amplitude and noise in the electrical signal [26]. However, the mechanical trapping has a trapping time not long enough to investigate the metabolic activity of bacteria and their interactions in large communities, so making also difficult their differentiation in the biofilm.

The aforementioned critical issues have been partly mitigated by Electrochemical Impedance Spectroscopy (*EIS*) [27–31]. In particular, device configurations based on interdigitated microelectrodes have been exploited for the analysis of the metabolic activity of bacteria and their differentiation [27]. This approach is characterized by a great penetration of the electric field in the biofilm, in contrast to the integrated optical devices where the distribution of the evanescent field of the optical resonant mode within the biofilm is limited [32], as demonstrated in the next sections. Therefore, the electrical behavior guarantees the capability of studying the biofilm evolution and the action of the antibiotics.

However, the lowest detection limit of *EIS* is around 10 CFU/mL [31], which means no high resolution down to single cells. Moreover, the detection time is longer compared to other trapping approaches, such as optical and mechanical ones, because, in case of low initial concentration, the growth of the cells is required before achieving a detectable change of impedance, which usually takes a few hours [33].

The resolution can be improved by different configurations of electrochemical biosensors, such as interdigitated nanoelectrodes, which compared to the bulk configuration of single electrodes allow the enhancement of the electric field in the device and then, a stronger interaction between the field and the bacteria [34]. However, if the resolution improvement is achieved at the expense of a shorter penetration length of the electric field, the analysis of the bacteria metabolism in a biofilm would be impossible.

From this brief overview on the techniques for the detection and analysis of bacteria, it is clear that a single interrogation technique is not sufficient for a full assessment of the antibiotic susceptibility on planktonic bacteria and, in particular, on bacterial biofilms. Multiple approaches should be used in parallel, leading to a multiparameter approach [35,36].

Here, we propose the design of a novel optoelectronic device based on an on-chip dual array of interdigitated micro- and nanoelectrodes that combines together optical and electrical techniques to monitor the growth of a biofilm and to analyze the effect of antibiotics on the bacteria. The optical approach allows the detection of few bacteria with a high spatial resolution to understand their interaction and which biological events are involved during the initial phase of the biofilm formation. The electrical approach allows the simultaneous study of the evolutionary phases of the bacterial biofilm and, by analyzing the impedance changes, the evaluation of the biofilm growth and maturation, and the efficacy of antibiotics for the disruption of the biofilm, which are useful in AMR studies.

3. Dual Array of Interdigitated Electrodes: Architecture and Operation

The dual array is formed by an Interdigitated Micro Electrodes (*IMEs*) section and an Interdigitated Nano Electrodes (*INEs*) section, which are both realized in Silicon-On-Sapphire (*SOS*) technology and fed by an *AC* voltage (see Figure 1a). In order to perform both optical and electrical measurements to study the properties of planktonic bacteria or a bacterial biofilm in the whole sensor area, the two sections are arranged adjacent, spaced from each other only by a few microns.

P-doping has been assumed for silicon, with an exponential decay of the electrical conductivity from the surface in depth with a drop more than three orders of magnitude in few tens of nanometers, as described in Section 4.1, reaching very low values of resistivity at the surface, and also strongly reducing the optical losses correlated to the dopant concentration [37]. The *INEs* section can be assumed as a top-illuminated subwavelength grating, whose cross-section is sketched in Figure 1, that supports the Guided Mode Resonance (*GMR*) effect [38]. It is correlated to the quasi-guided modes or leaky modes of the structure, as shown by the grating in Figure 1a. The grating acts as a waveguiding layer in the x-y plane, where the input light excites quasi-guided or leaky modes that coherently scatter at each interface of the grating. Furthermore, leaky modes scatter power downwards, along the vertical direction perpendicular to the grating (z-axis).

By properly engineering the grating features, such as the period, the refractive indices, and the angle of incidence, the interference between the transmitted light and the downward wave scattered by the leaky mode could generate reflected light along the negative direction of the z-axis with maximum amplitude at resonance [39,40].

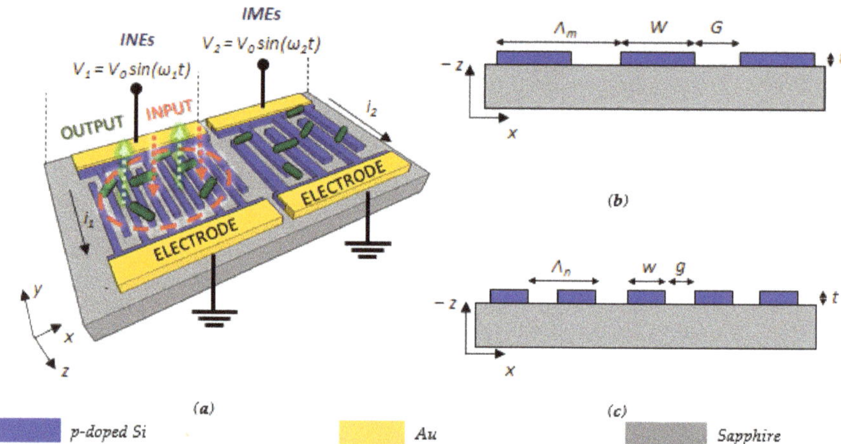

Figure 1. (**a**) Configuration of the dual array of interdigitated micro- and nanoelectrodes in SOS technology for the electrical detection of bacteria (in green) growth and metabolism. The section with nanoelectrodes represents an optical sub-wavelength grating resonating at a specific wavelength when top illuminated, so enabling a simultaneous optical and electrical detection of bacteria deposition; (**b**) Interdigitated Micro Electrodes (*IMEs*), where *W* is the width of the silicon layer (*W >> w*), *G* is the gap between electrodes (*G >> g*); and Λ_m is the *IMEs* period; (**c**) Interdigitated Nano Electrodes (*INEs*), where *w* is the width of the silicon layer, *g* is the gap between electrodes; and Λ_n is the *INEs* period. The output signal of the Guided Mode Resonance (GMR) structure consists of the reflected optical signal (green arrows) by illuminating the optical section with *TE*-polarized light (red arrows).

The structure exploiting the *GMR* effect has been designed to obtain a resonance condition for $\lambda > 800$ nm, where the absorption losses of silicon decrease [41], with the aim of achieving a higher extinction ratio and higher *Q*-factor, together with a strong energy confinement close to the surface to enhance the light interaction with the bacteria. The strong refractive index contrast between silicon and the surrounding medium allows a high confinement of the electromagnetic field at the sensor surface, enabling the use of the *INEs* section for hyperspectral imaging technique, as described in detail in [42], which allows the refractive index imaging, thus localizing objects on the grating by detecting the spatial resonance distribution. An inverted microscope configuration can be used to characterize the sensor, with the light source illuminating from the top and the reflected signal collected from the same side of the setup by a conventional CMOS camera.

In order to work at the bacteria scale, we assume as a best compromise in terms of resolution and the large field of view an optical setup with an area of 1 mm^2 and a spatial resolution down to few microns. This allows having a large area but still exploiting the advantages of near-field optics, in particular in terms of strong resolution. This approach has been experimentally validated in the literature with simple and feasible systems [42]. This behavior is also useful to obtain additional information about the first stage of the bacteria infection, when the bacteria cells enrich before they start to interact with each other to form the biofilm. During the biofilm formation, the cells produce extracellular polymeric substances (*EPS*), creating the surrounding matrix to protect the bacteria. The main *EPS* components are polysaccharides, proteins, lipids, and *DNA* with dimensions much smaller than the bacterial cells [43]. The real-time detection of the biofilm formation with label-free techniques is very challenging, mainly because the matrix generation begins when the first layers of bacteria close to the sensor surface are packed and arranged in large communities. Under this condition, the monitoring of the chemical and biological processes in these communities is difficult because the propagation length of the evanescent field, of the order of few hundreds of nanometers, corresponds only to the first layer of bacteria. Therefore, the optical approach is efficient for analyzing the first stages of the biofilm formation when bacteria begin to form colonies, while a different approach should be investigated to clearly define the presence of a biofilm with its extracellular matrix. To meet this latter requirement, the *INEs* section has been designed to allow simultaneous optical and electrical measurements. In fact, several pairs of nanoelectrodes can be realized in an interdigitated configuration, by connecting in turn the doped silicon structures to two different metal electrodes, as shown in Figure 1c, so forming several capacitors at the nanoscale. This system of interdigitated electrodes can allow easy monitoring of the large change of the *PH* and the electrical properties of the solution induced by the secretion of proteins, *DNA*, and other *EPS* components when the bacteria colonies start to create the extracellular matrix [44]. In particular, the capacitance of the system changes when the bacteria start to grow because of the strong interaction between the electric field and the bacteria. Low-frequency values are usually used to detect changes of the system capacitance, because under this condition, the bacterial membrane behaves as a barrier to prevent the penetration of the electric field into the cytoplasm of the bacteria, which usually has a much higher conductivity, in order to make more evident any change of the impedance in proximity of the electrodes. The *INE* section supporting the *GMR* effect guarantees the confinement of the electric field close to its top surface [34], where the bacteria grow and the chemical processes happen, with an improved sensitivity compared to other electrical approaches with different systems and configurations.

Therefore, the *INEs* provide high spatial resolution for imaging few bacteria through an optical approach, to investigate the bacteria interaction to form colonies, and an electrical approach to analyze the biofilm evolution, due to the presence of *EPS* components that change the electrical properties of the solution, and, therefore, interfere with the electric field distribution. This interaction induces a change of the capacitance C_i of the nanocapacitor formed by each pair of nanoelectrodes, which depends on the electrical properties of the surrounding medium and the electrode geometry [44]. The net capacitance of the system, C_{tot}, is given by the sum of the single capacitances C_i in parallel combination.

A voltage of few millivolts and low frequency have been assumed, to have a low power interacting with bacteria with consequent strongly reduced risks, such as bacteria membrane collapses or changes of their metabolic state. An *AC* signal at low frequency, i.e., $f_1 \approx 100$ Hz, is necessary to detect any change of the system capacitance [45]. The net impedance of the system is given by $Z = 1/j\omega C_{tot} = 1/j\omega N C_i$, where N is the number of nanocapacitors in the interdigitated configuration. This corresponds to a maximum value of current $i_{max} = V_0 / |Z| = \omega V_0 N C_i$, which can be increased with the number of pairs of electrodes, leading to the increase in the Signal-to-Noise Ratio (*SNR*) and the sensitivity.

Since a higher sensitivity also corresponds to a faster saturation of the impedance change, due to the strong confinement of the electric field, which is not affected by the impedance changes in the upper layers of the biofilm, the main limitation of the *INEs* is given by the narrow dynamic range [46]. To monitor the biofilm upper layers, a great penetration depth of the electric field is needed, with a resulting large dynamic range and a lower sensitivity. For this reason, the sensor also includes an *IMEs* section placed next to the *INEs* structure (see Figure 1a). Since the penetration depth L of the electric field is proportional to the gap G and to the width W of the electrodes ($L–W$), features at the microscale lead to a large penetration depth and dynamic range [47]. In particular, by assuming $W > 10$ μm, any change of the metabolic state of the biofilm can be monitored as an impedance change. To improve the *SNR*, an *IMEs* driving voltage frequency f_2 larger than 1 kHz can be used, where the biofilm shows resistive behavior. According to the above-mentioned net impedance equation, an increase in the electrodes' driving voltage leads to a decrease in the impedance Z, allowing improvement in the accuracy of the detected changes in the biofilm with respect to the *INEs* section.

Although *INEs*- and *IMEs*-based biosensors have been already widely investigated, see as examples [48,49], we note that the main novelty of the proposed optoelectronic device is the combination of both optical and electrical approaches to perform on the same platform and at the same time the efficient monitoring of the bacteria growth and the analysis of the resulting biofilm under antibiotics treatment. According to the above, it can be said that the high sensitivity is provided by *INEs* through a strong confinement of the electric field close to the top silicon surface, while the *IMEs* allow analyzing the growth and the maturation of the biofilm and studying its full or partial disruption induced by the antibiotic treatment for a complete, accurate, and real-time monitoring of the biofilm properties and analysis of *AMR* to specific antibiotics. The driving voltages V_1 and V_2 for *INEs* and *IMEs*, respectively, can be the same to simplify the setup of the system, while an *IMEs* driving voltage frequency f_2 larger than the *INEs* one is preferred to improve the SNR.

The system configuration could be realized following typical fabrication processes. For example, a lithographic process followed by reactive-ion etching can be used to define both the *INEs* and *IMEs* structures, while photolithography followed by metal evaporation and lift-off can be applied for the metal pads (Figure 1).

4. Design of the Dual Array of Interdigitated Electrodes

4.1. Design of the INEs for Optical and Electrical Measurements

The optical section, as already mentioned, consists of a *GMR* structure with a sub-wavelength grating in SOS technology. A doped silicon was assumed for doing electrical measurements, with a dopant concentration $N_D = 10^{21}$ cm^{-3} at the surface and an exponential drop-off to $N_D = 10^{18}$ cm^{-3} within 20 nm, which can be achieved by thermal diffusion doping [37,50]. This doping profile allows minimizing the optical losses, without any worsening of electrical performance. The geometrical features of the silicon subwavelength grating have been designed to enhance both electrical and optical performance. A resonance condition around $\lambda = \lambda_{res} \approx 850$ nm is required to minimize optical losses due to the material absorption typical for silicon at lower wavelengths. To fulfill all the requirements, including fabrication tolerances, we have determined a thickness $t = 270$ nm, a period $\Lambda_n = 440$ nm, and a fill factor $FF = 0.5$, corresponding to $w = g = 220$ nm (see Figure 2a). Furthermore, since the grating strength, and then the Q-factor, increases with the number of periods, thousands of periods have been assumed, without affecting the ability to detect the bacteria with a resolution of a few microns.

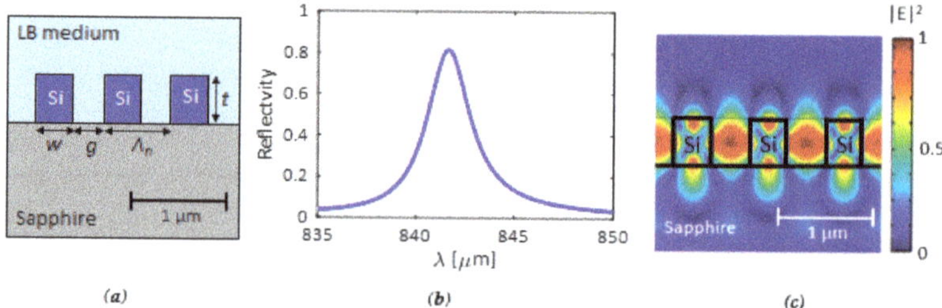

Figure 2. (a) Configuration of the *GMR* structure (*LB*: Lysogeny Broth); (b) Reflection spectrum; (c) Energy confinement at the resonance wavelength $\lambda \approx 841.5$ nm.

The reflection spectrum of the grating has been calculated by the 3D Finite Element Method (*FEM*), assuming top out-of-plane excitation with *TE*-polarized and plane-wave collimated light (the electric field is oriented perpendicularly to the grating period direction).

Reflection spectrum and mode distribution at the resonance frequency are reported in Figure 2b,c, respectively.

The Lysogeny Broth (*LB*) ($n_{LB} = 1.333 + 5 \times 10^{-7}$ i) has been assumed as the surrounding medium, which is a typical medium for bacteria, and the substrate is sapphire with $n_{sub} = 1.732$. A resonance condition at $\lambda_{res} \approx 841.5$ nm has been calculated. At λ_{res}, with the aforementioned doping profile, a refractive index of silicon $n_{Si} = 2.76 + 0.06$ i at the surface results, which increases up to $n_{Si} = 3.648 + 4 \times 10^{-3}$ i at 20 nm far from the surface. The resonance shows an amplitude higher than 0.8, which is normalized with respect to the input power, and a Full Width at Half Maximum (*FWHM*) = 2.12 nm, which corresponds to $Q \approx 400$. The proposed configuration represents the best compromise to achieve a high Q-factor, large modal confinement, and high resonance amplitude, which is useful for hyperspectral imaging and sensing.

The operation of the *GMR* structure in the presence of bacteria has been simulated by assuming a uniform and homogeneous layer of bacteria with a thickness t_{layer} of the order of hundreds of nm and refractive index n_{bac}. For example, the *Escherichia coli* bacterium was considered, with a diameter t_{bac} of 500 nm, a length l_{bac} of 2 μm, and a refractive index $n_{bac} = 1.388$ [51]. Since the length of the bacteria l_{bac} is larger than the gap size g, the penetration of bacteria in the grooves is not allowed. Therefore, the grooves have been assumed to be filled by *LB* medium. The comparison between the 1D silicon *GMR* surrounded by *LB* only and the same structure with the presence of bacteria in solution is shown in Figure 3. A resonance shift $\Delta\lambda_{res}$ of about 0.9 nm in the case of the bacteria layer on the *GMR* structure, with a reflection change of about 0.02, was evaluated. This behavior confirms a high resolution in detecting the presence of the bacteria, even when the surface is not fully covered.

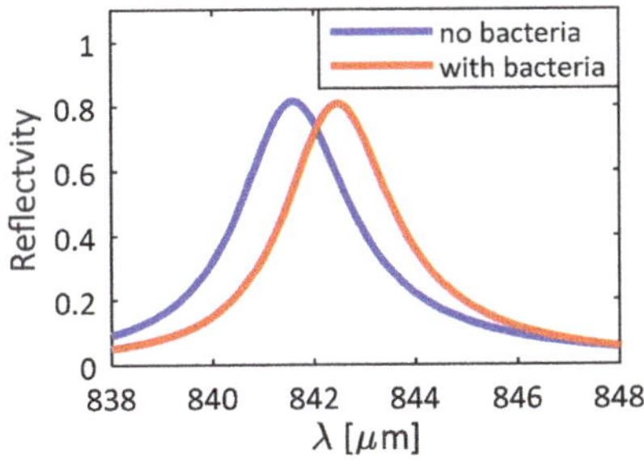

Figure 3. Reflection spectra for the case without bacteria (blue curve) and with bacteria (red curve) on the *GMR* structure. Spectra calculated by 3D FEM approach.

Since the evanescent field of the optical mode extends and interacts with the biofilm for a few hundred nanometers ($<<t_{bac}$ = 500 nm), beyond the first layer of bacteria, the optical response results are insensitive to an increase in the layer thickness, as confirmed by FEM simulations where the behavior of the reflected signal is the same for $t_{layer} \geq t_{bac}$ with negligible resonance shifts. However, during the biofilm formation, a clearer resonance shift is also expected because of the release of small particles and molecules, possibly in the grooves, which would further affect the effective index. As an example, in the extreme case of filling the grooves with biomolecules secreted by the *Escherichia coli* bacteria (for which we assume the same refractive as for the bacteria, $n_{bac} = 1.388$), 3D *FEM* simulations confirm a maximum resonance shift up to 7.5 nm. Therefore, in the presence of a biofilm, a final value of the resonance shift in the range 0.9 nm $< \Delta\lambda_{res} <$ 7.5 nm is expected.

When investigating the resonance shift in an array of several pixels, the resonance map for each pixel can give information about the position of bacteria and the size of colonies.

However, due to the short penetration length of the evanescent field in the surrounding medium, this approach cannot provide a very accurate analysis about the formation of a biofilm and its maturation. This limitation justifies the choice of a more complex biosensing platform by combining optical measurements with electrical ones. Hence, the same interdigitated silicon nanoelectrodes have also been used for electrical measurements, as shown in Figure 1a, where an applied voltage $V_0 = 10$ mV enables the current flow i_1. In order to not interfere with the optical response, the metal pads of this structure are far enough from the region used for optical measurements. An electrode length of about 1 mm fulfills this requirement, also avoiding any power absorption of the medal pads, with a consequent reduction of the interacting optical power. The main goal of the *INE* structure is detecting the presence of bacteria and the first stage of bacterial biofilm formation. The maximum sensitivity can be reached by evaluating the changes of the system capacitance, where the capacitance is given by $C = \varepsilon A / d$, with ε being the relative permittivity of the surrounding medium, A being the area of the electrodes, and d being the distance between the silicon surface and the charged particles released by the bacteria [52]. Electrical measurements require strong confinement of the electric field at the surface of silicon, which is achievable with a typical frequency $f_1 \approx 100$ Hz. The relative permittivity of *LB* medium at $f_1 = 100$ Hz has been assumed to be equal to $\varepsilon = 60$, and the electrical conductivity $\sigma_{LB} = 0.754$ S/m. As the first step, the electrical behavior of the *INEs* has been evaluated by 2D *FEM* simulations without the presence of bacteria to observe the electric field distribution in the proximity of the electrodes. The grating length is much larger than the grating period and the electrodes features; therefore, we have assumed a 2D simulation as an optimal approximation of the real case of study. Figure 4 shows the distribution of the current density J_1 [A/m^2] without bacteria. As expected, the energy decreases as a function of the distance along the z-axis from the electrodes and increases with a peak of the energy density ($\approx 7 \times 10^{-4}$ A/m^2) at the silicon ridges. This behavior confirms the suitability of the electrical measurements to monitor the biofilm at the silicon surface.

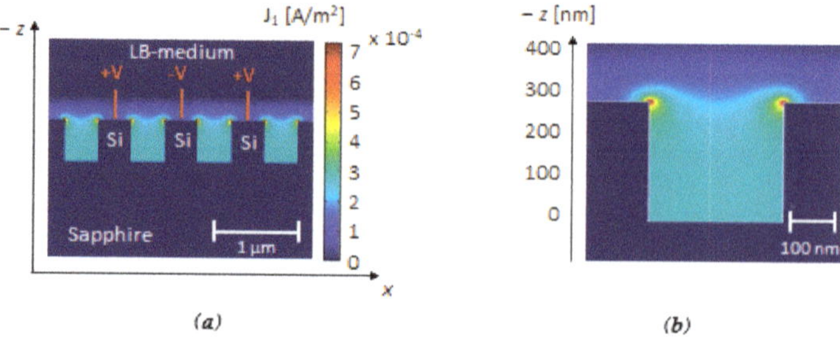

Figure 4. (**a**) Current density J_1 [A/m^2] distribution along the *x*-axis in the middle of the *INEs* structure with $f_1 = 100$ Hz, $V_0 = 10$ mV, $\Lambda_n = 440$ nm, and $FF = 0.5$; (**b**) Focus at the surface of the electrodes with the energy confinement in few hundreds of nanometers. Plots calculated by 2D FEM approach.

The total capacitance of the *INE* structure is directly proportional to the number of pairs of electrodes. A number of couples $N = 2000$ and a length of electrodes of the order of mm have been assumed in the model, which corresponds to a total width of the system of $N \cdot \Lambda = 2000 \times 0.44$ µm = 880 µm. The footprint of the order of mm^2 (=880 µm $\times \approx$1 mm) also guarantees a large area of the *INEs* section optimizing the optical reflection and obtaining more information for large bacterial colonies. The capacitance of a system with interdigitated electrodes is [53]:

$$C = \frac{N \cdot Q}{2V_0} = N \cdot \frac{2}{V_0{}^2} E = N \cdot \frac{2}{V_0{}^2} \int_\Omega W_e d\Omega \tag{1}$$

where Ω is the surrounding area of the electrodes close to their surface [m^2], E is the energy for each pair of electrodes [J], and W_e is the electric energy density [J/m^2]. Assuming $V_0 = 10$ mV and $f_1 = 100$ Hz, the energy E is equal to 31 fJ, which corresponds to a total system capacitance $C = 1.24$ µF. In the low-frequency regime ($f_1 = 100$ Hz), the system behavior is capacitive, with a corresponding impedance $Z_c = 1/j\omega C$, decreasing as C increases. $N = 2000$ corresponds to $Z_c = 1.2$ kΩ. The maximum value of the current $i_{1,max}$ is given by [54]:

$$i_{1,max} = max\left(C\frac{\partial V}{\partial t}\right) = max\left(C\frac{\partial(V_0 \sin(\omega t))}{\partial t}\right) = 2\pi f_1 C V_0 = 7.79 \text{ µA.} \tag{2}$$

To simulate the biofilm formation and maturation in the *LB*, the well-established Maxwell Mixture Theory (*MMT*) [55] has been used to define the electric properties of the biofilm. In particular, the *MMT* method assumes the biofilm as a compound of uniformly distributed spherical objects as the bacterial cells, which are covered by a shell to mimic the external membranes, forming the so-called Extracellular Matrix (*ECM*) [46]. With these assumptions, the dielectric permittivity and electrical conductivity of the biofilm can be theoretically estimated. In particular, the relative permittivity of the region with the biofilm $\varepsilon^*_{biofilm}(\omega)$ is given by [55]:

$$\varepsilon^*_{biofilm}(\omega) = \varepsilon^*_{ECM}(\omega)\frac{2(1-\varphi)\varepsilon^*_{med} + (1+2\varphi)\varepsilon^*_{eq}(\omega)}{(2+\varphi)\varepsilon^*_{med} + (1-\varphi)\varepsilon^*_{eq}(\omega)} \tag{3}$$

where φ is the fractional volume of the bacterial cells in the *ECM* that has been assumed equal to 30% following a conservative approach [45], ε^*_{MED} is the complex permittivity of the solution where the bacteria are immersed, and $\varepsilon^*_{eq}(\omega)$ is the equivalent complex dielectric constant of the bacteria, expressed as [55]:

$$\varepsilon^*_{eq}(\omega) = \varepsilon^*_{mem}(\omega)\frac{2(1-\theta)\varepsilon^*_{mem} + (1+2\theta)\varepsilon^*_{cyt}(\omega)}{(2+\theta)\varepsilon^*_{mem} + (1-\theta)\varepsilon^*_{cyt}(\omega)} \tag{4}$$

with the complex permittivity of the *ECM* $\varepsilon^*_{ECM}(\omega) = \varepsilon_{r\_ECM} + \sigma_{ECM}/(j\varepsilon_0\omega)$, the complex permittivity of the bacterial cell membrane $\varepsilon^*_{MEM}(\omega) = \varepsilon_{r\_MEM} + \sigma_{MEM}/(j\varepsilon_0\omega)$, and the permittivity of the bacterial cytoplasm $\varepsilon^*_{CYT}(\omega) = \varepsilon_{r\_CYT} + \sigma_{CYT}/(j\varepsilon_0\omega)$. Moreover, $\theta = (R/(R + d))$, with R and d being the radius of the bacteria and the thickness of the external membrane, respectively, and the parameters ε_0 and ω being the dielectric permittivity in vacuum and the angular frequency of the applied signal, respectively. The conductivity of the biofilm is calculated as $\varepsilon_{biofilm} = \varepsilon_{r\_biofilm} + \sigma_{biofilm}/(j\varepsilon_0\omega)$. According to the *MMT* theory, a negligible change would be obtained with a model that assumes bacteria with an ellipsoidal shape instead of a spherical one. The parameters values used in the numerical model to implement the *MMT* are reported in Table 1. The electrical properties are derived by experimental measurements reported in the literature [46,56]

Table 1. Electrical properties of the parameters used in the *MMT* [46,56].

	Conductivity [S/m]	Relative Permittivity [a.u.]
Cytoplasm (*CYT*)	0.220	100
Membrane (*MEM*)	10^{-7}	10.8
Extracellular Matrix (*ECM*)	0.680	60
Lysogeny Broth (*LB*)	0.754	60

An initial value of the capacitance $C_0 = 1.24$ μF has been simulated by assuming only *LB* medium above the nanoelectrodes. As already assumed in the optical analysis, the presence of a biofilm layer with a thickness of 1 μm, above the nanoelectrodes, was considered. Under this condition, the capacitance becomes $C' = 1.41$ μF, the capacitance relative change is $\Delta C/C_0 = (C' - C_0)/C_0{\sim}14\%$, and the maximum value of current is equal to $i_{1,max} = 8.85$ μA. This performance confirms the high sensitivity of the nanoelectrodes because a change of the current values of about 14% is obtained when a single layer of bacteria is placed on top of the nanoelectrodes (thickness = 1 μm). This behavior is strictly correlated to the strong confinement of the electric field at the surface of the electrodes and represents a significant advantage with respect to optical measurements in terms of sensitivity to the biofilm formation and maturation. However, the electrical measurement takes into account only an average change of the surrounding medium with a spatial resolution > 1 μm, while the optical approach provides the spatial distribution of bacteria along the grating with a much higher resolution, of the order of hundreds of nm, so demonstrating the strong complementarity of the methods. Since the strong electric field confinement of the *INE* structure causes the saturation of the impedance value, even with a single layer of bacteria in the biofilm, negligible changes of the impedance for a thicker layer of biofilm were confirmed by *FEM* simulations. This restriction justifies the combination of the *INE* and *IME* structures in order to also detect a thicker biofilm for the study of the maturation phase and for the analysis of a possible biofilm disruption by using antibiotics whose results are challenging only with the *INE* structure.

4.2. Design of the IMEs for the Detection of Biofilm Maturation or Disruption

The same thickness and doping distribution of the *INEs* described in Section 4.1 were assumed for the configuration of the *IMEs* to realize the proposed array with a single manufacturing process. As above introduced, the *IMEs* function is providing an accurate analysis of the biofilm maturation and its possible disruption through an interaction of the electric field distribution with the upper layers of the bacterial biofilm. A design different from *INEs*, in terms of width W, gap G between the electrodes, and period Λ_m (see Figure 1c) is required. The electric field distribution for different values of width W is in Figure 5, for different values of W ($W = 5$ μm, 10 μm, and 15 μm with $W = G$). The numerical results confirm that a larger value of W and G allows confining the electric field farther away from the electrodes, as required for the *IME* structure, at the expense of a decrease in the current density J_2 [A/m^2] due to a worsening of the related capacitance value.

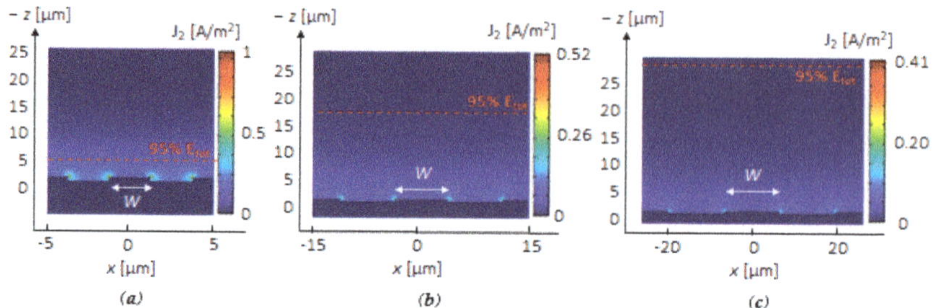

Figure 5. Current density J_2 [A/m^2] distribution in the *IME* structures with a value of width $W = G$ of 5 μm (**a**), 15 μm (**b**), and 25 μm (**c**). The current values have been normalized to the maximum value of current calculated for $W = 5$ μm. The dotted red line represents the distance from the electrodes, where 95% of the total energy is confined. Plots calculated by 2D FEM approach.

The performance of the *IME* configuration without the bacterial biofilm was defined by *FEM* simulations for different values of the width, assuming a number of electrodes $N = 100$, aiming at preserving the device compactness. Figure 6 shows the change of the capacitance for different values of W (assuming $G = W$) with respect to the capacitance C_0 calculated when the *IMEs* are not covered by the biofilm layer.

The same model based on the *MMT* has been used to define the electrical performance for *IMEs*, as already proposed for the nanoelectrodes. The presence of multiple layers of bacteria has been considered for the biofilm, with each layer of bacteria assumed with a thickness of 1 μm.

As expected, the capacitance changes quickly for small gap values due to the stronger energy confinement, but this also corresponds to a more evident nonlinear behavior for a thickness of the biofilm of few microns (see Figure 6a), which makes the rigorous analysis of the biofilm behavior challenging, even with several layers of bacteria.

For example, a value of $G = W = 5$ μm presents a nonlinear behavior up to 5 μm, where the capacitance reaches its saturation value. For a biofilm thickness larger than 5 μm, the changes of impedance are negligible, making impossible the analysis of the upper biofilm layers. On the contrary, a larger value of G provides a more evident linear behavior of the capacitance change with respect to the thickness of the biofilm at the expense of less sensitivity. In fact, for $G = W = 100$ μm, a linear behavior of the capacitance has been observed for a thickness of the biofilm of at least 15 μm. However, an impedance change of about 10% has been calculated for a thickness of the biofilm of 10 μm, while the impedance variation goes down to 5% with a thickness of 5 μm, which is six times lower than the performance obtained with $G = 5$ μm for the same biofilm thickness.

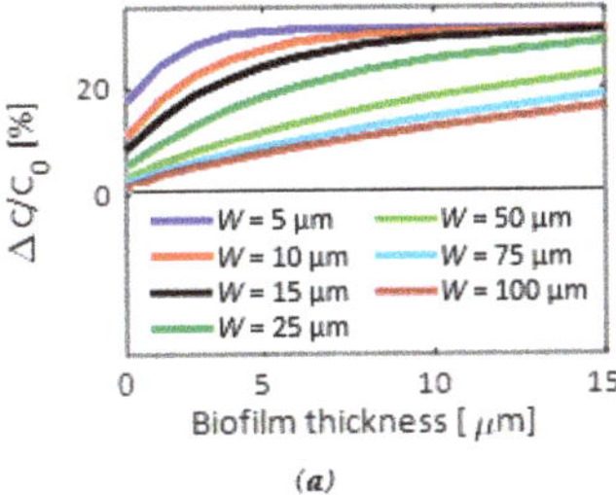
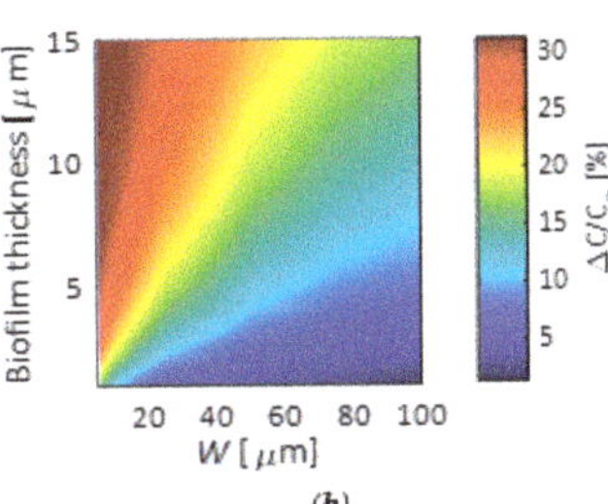

(a)   (b)

Figure 6. (a) Change of the capacitance $\Delta C/C_0 = (C' - C_0)/C_0$ in the *IME* structure for different values of thickness of the bacterial biofilm as a function of the electrode width ($W = G$); (b) Change of the capacitance $\Delta C/C_0$ as a function of the electrode width calculated for different values of the biofilm thickness.

Figure 6b confirms the nonlinear trend of the capacitance for smaller values of G, which becomes negligible with a larger W, in addition to a narrower dynamic range of the relative capacitance change $\Delta C/C_0$ that also corresponds to a lower sensitivity. The conditions $G = W = 50$ μm, and then, $\Lambda_m = 100$ μm, have been chosen as the best compromise in terms of linearity and sensitivity, obtaining an impedance change up to 20% with a biofilm thickness of 15 μm. Since the detection of the changes of capacitance at the interface of the electrodes is not necessary, as required instead for the *INEs*, a frequency f_2 of 1 kHz has been assumed for the simulation of the current (i_2 in Figure 1a) changes with different thickness of the biofilm in order to probe electrical changes in the bulk solution [57].

The electrical performance by varying the biofilm thickness is reported in Figure 7. A change of the current i_2 from $i_2 = 1.76$ up to 2.09 μA with a quasi-linear behavior has been calculated, enabling the detection of the current change for each layer of biofilm that confirms the ability to easily detect both the bacterial biofilm growth and maturation, together with its possible disruption caused by the action of the antibiotics. A similar behavior represents a significant improvement for *AMR* because an efficient, accurate, and real-time analysis of the bacteria interaction and useful information about their life in community and colonies can be achieved. Furthermore, the electrical measurements of the *IMEs* ensure an accurate analysis of the metabolic state of the biofilm during the whole process from the formation to the maturation and, possibly, the disruption for specific antibiotics.

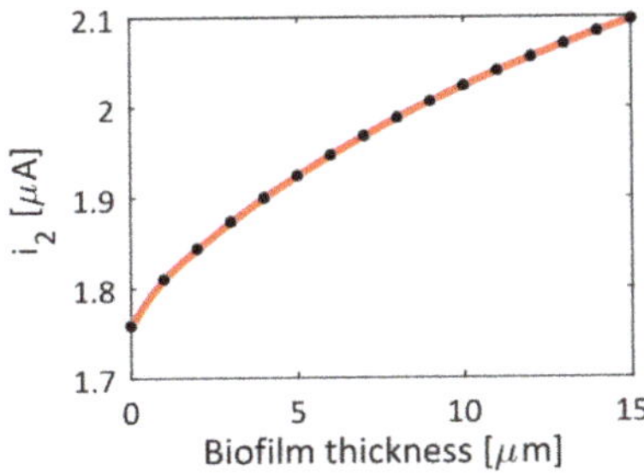

Figure 7. Current i_2 [μA] in the *IME* structure with $N = 100$, $f_2 = 1$ kHz, and $G = W = 50$ μm for different values of biofilm thickness. Plots calculated by 2D FEM approach.

5. Discussions and Conclusions

An innovative label-free biosensing platform based on a dual array of interdigitated electrodes for simultaneous optical and electrical measurements has been proposed for the analysis of bacteria and their interaction. A multiparametric approach for the *INEs* section offers high sensitivity to detect a very low concentration of planktonic bacteria and possibly down to a single cell, also ensuring the following of the biofilm formation in the initial stage. In addition, the electrical performance of the *IMEs* enables the monitoring of the growth and the maturation of the biofilm to possibly investigate the efficiency of the antibiotics. The overall performance, related to the merging of the micro and nano scale, outperforms the competitive technologies, whose operation is limited to monitoring a few biofilm formation stages. Therefore, the main advantage of the proposed system is the capability to detect and monitor in real time a biofilm, also analyzing its metabolic state and evolution phase. In its proof-of-concept form, the proposed optoelectronic device has been used for the monitoring of *Escherichia coli*-based biofilm, but the device could also be investigated to analyze biofilms formed by different strains and species of bacteria. Therefore, the proposed detection method's results are very promising due to high sensitivity, low-cost fabrication, and real-time operation, paving the way to a real-time and cost-effective solution to counteract the AMR phenomenon.

Author Contributions: Conceptualization, G.B., D.C., M.N.A. and C.C.; methodology, G.B. and D.C.; software, G.B. and D.C.; formal analysis, G.B. and D.C.; investigation, G.B. and D.C.; writing—original draft preparation, G.B.; writing—review and editing, D.C., M.N.A. and C.C.; supervision, M.N.A. and C.C.; funding acquisition, G.B. and C.C. All authors have read and agreed to the published version of the manuscript.

Funding: This research was funded by POR Puglia FESR FSR 2014–2020–Action 10.4–"Research for Innovation" (REFIN) Initiative.

Institutional Review Board Statement: Not applicable.

Informed Consent Statement: Not applicable.

Data Availability Statement: Not applicable.

Conflicts of Interest: The authors declare no conflict of interest.

References

1. World Health Organization. *World Health Statistics 2020*; World Health Organization: Geneva, Switzerland, 2020; Available online: https://apps.who.int/iris/bitstream/handle/10665/332070/9789240005105-eng.pdf (accessed on 18 September 2021).
2. World Health Organization. *WHO Estimates of the Global Burden of Foodborne Diseases: Foodborne Disease Burden Epidemiology Reference Group 2007–2015*; World Health Organization: Geneva, Switzerland, 2015; Available online: https://www.who.int/publications/i/item/9789241565165 (accessed on 18 September 2021).
3. Dadgostar, P. Antimicrobial resistance: Implications and costs. *Infect. Drug Resist.* **2019**, *12*, 3903–3910. [CrossRef] [PubMed]
4. Stewart, P.S.; Costerton, J.W. Antibiotic resistance of bacteria in biofilm. *Lancet* **2001**, *358*, 135–138. [CrossRef]
5. Andersson, D.I.; Hughes, D. Antibiotic resistance and its cost: Is it possible to reverse resistance? *Nat. Rev. Microbiol.* **2010**, *8*, 260–271. [CrossRef]
6. Chang, H.H.; Cohe, T.; Grad, Y.H.; Hanage, W.P.; O'Brien, T.F.; Lipsitch, M. Origin and proliferation of multiple-drug resistance in bacterial pathogens. *Microbiol. Mol. Biol. Rev.* **2015**, *79*, 101–116. [CrossRef]
7. Review on Antimicrobial Resistance. *Antimicrobial Resistance: Tackling a Crisis for the Health and Wealth of Nations*; Review on Antimicrobial Resistance: London, UK, 2014.
8. Flemming, H.C.; Wingender, J.; Szewzyk, U.; Steinberg, P.; Rice, S.A.; Kjelleberg, S. Biofilms: An emergent form of bacterial life. *Nat. Rev. Microbiol.* **2016**, *14*, 563–575. [CrossRef] [PubMed]
9. Costerton, J.W.; Stewart, P.S.; Greenberg, E.P. Bacterial biofilms: A common cause of persistent infections. *Science* **1999**, *284*, 1318–1322. [CrossRef] [PubMed]
10. Flemming, H.C.; Wingender, J. The biofilm matrix. *Nat. Rev. Microbiol.* **2018**, *8*, 623–633. [CrossRef]
11. De la Fuente-Nunez, C.; Reffuveille, F.; Fernandez, L.; Hancock, R.E.W. Bacterial biofilm development as a multicellular adaption: Antibiot resistance and new therapeutic strategies. *Curr. Opin. Microbiol.* **2013**, *16*, 580–589. [CrossRef]
12. Olson, M.E.; Ceri, H.; Morck, D.W.; Buret, A.G.; Read, R.R. Biofilm bacteria: Formation and comparative susceptibility to antibiotics. *Can. J. Vet. Res.* **2002**, *66*, 86–92.

13. Arabski, M.; Konieczna, I.; Tusinska, E.; Wasik, S.; Relich, I.; Zajac, K.; Kaminski, Z.J.; Kaca, W. The use of lysozyme modified with fluorescein for the detection of Gram-positive bacteria. *Microbiol. Res.* **2015**, *170*, 242–247. [CrossRef]

14. Gao, X.; Guo, M.; Zhang, Z.; Shen, P.; Yang, Z.; Zhang, N. Baicalin promotes the bacteriostatic activity of lysozyme on *S. aureus* in mammary glands and neutrophilic granulocytes in mice. *Oncotarget* **2017**, *8*, 19894. [PubMed]

15. Dickson, R.P.; Singer, B.H.; Newstead, M.W.; Falkowski, N.R.; Erb-Downward, J.R.; Standiford, T.J.; Huffnagle, G.B. Enrichment of the lung microbiome with gut bacteria in sepsis and the acute respiratory distress syndrome. *Nat. Microbiol.* **2016**, *1*, 16113. [CrossRef] [PubMed]

16. Bassetti, M.; Poulakou, G.; Ruppe, E.; Bouza, E.; Van Hal, S.J.; Brink, A. Antimicrobial resistance in the next 30 years, humankind, bugs and drugs: A visionary approach. *Intensive Care Med.* **2017**, *43*, 1464–1475. [CrossRef] [PubMed]

17. Ciminelli, C.; Conteduca, D.; Dell'Olio, F.; Armenise, M.N. Design of an optical trapping device based on an ultra-high Q/V resonant structure. *IEEE Photonics J.* **2014**, *6*, 1–16. [CrossRef]

18. Conteduca, D.; Reardon, C.; Scullion, M.G.; Dell'Olio, F.; Armenise, M.N.; Krauss, T.F.; Ciminelli, C. Ultra-high Q/V hybrid cavity for strong light-matter interaction. *APL Photonics* **2017**, *2*, 086101. [CrossRef]

19. Baron, V.O.; Chen, M.; Clark, S.O.; Williams, A.; Dholakia, K.; Gillespie, S.H. Detecting Phenotiypically Resistant Mycobacterium tuberculosis using wavelength modulated Raman spectroscopy. In *Antibiotic Resistance Protocols. Methods in Molecular Biology*; Humana Press: New York, NY, USA, 2018; pp. 41–50.

20. Righini, M.; Ghenuche, P.; Cherukulappurath, S.; Myroshnychenko, V.; Garcia de Abajo, F.J.; Quidant, R. Nano-optical trapping of rayleigh particles and *Escherichia coli* bacteria with resonant optical antennas. *Nano Lett.* **2009**, *9*, 3387–3391. [CrossRef]

21. Therisod, R.; Tardif, M.; Marcoux, P.R.; Picard, E.; Jager, J.B.; Hadji, E.; Peyrade, D.; Houdre, R. Gram-type differentiation of bacteria with 2D hollow photonic crystal cavities. *Appl. Phys. Lett.* **2018**, *113*, 111101. [CrossRef]

22. Conteduca, D.; Brunetti, G.; Dell'Olio, F.; Armenise, M.N.; Krauss, T.F.; Ciminelli, C. Monitoring of individual bacteria using electro-photonic traps. *Biomed. Opt. Express* **2019**, *10*, 3463–3471. [CrossRef]

23. Wang, Y.; Reardon, C.P.; Read, N.; Thorpe, S.; Evans, A.; Todd, N.; Van Der Woude, M.; Krauss, T.F. Attachment and antibiotic response of early-stage biofilms studied using resonant hyperspectral imaging. *NPJ Biofilms Microbiomes* **2020**, *6*, 1–7. [CrossRef]

24. Mu, X.; Zheng, W.; Sun, J.; Zhang, W.; Jiang, X. Microfluidics for manipulating cells. *Small* **2013**, *9*, 9–21. [CrossRef]

25. Etayash, H.; Khan, M.F.; Kaur, K.; Thundat, T. Microfluidic cantilever detects bacteria and measures their susceptibility to antibiotics in small confined volumes. *Nat. Commun.* **2016**, *7*, 1–9. [CrossRef]

26. Li, M.; Xi, N.; Wang, Y.; Liu, L. Advances in atomic force microscopy for single-cell analysis. *Nano Res.* **2019**, *12*, 703–718. [CrossRef]

27. Kim, S.; Yu, G.; Kim, T.; Shin, K.; Yoon, J. Rapid bacterial detection with an interdigitated array electrode by electrochemical impedance spectroscopy. *Electrochim. Acta* **2012**, *82*, 126–131. [CrossRef]

28. Leva-Bueno, J.; Peyman, S.A.; Millner, P.A. A review on impedimetric immunosensors for pathogen and biomarker detection. *Med. Microbiol. Immunol.* **2020**, *209*, 343–362. [CrossRef] [PubMed]

29. Cesewski, E.; Johnson, B.N. Electrochemical biosensors for pathogen detection. *Biosens. Bioelectron.* **2020**, *159*, 112214. [CrossRef] [PubMed]

30. Rashed, M.Z.; Kopechek, J.A.; Priddy, M.C.; Hamorsky, K.T.; Palmer, K.E.; Mittal, N.; Valdez, J.; Flynn, J.; Williams, S.J. Rapid detection of SARS-CoV-2 antibodies using electrochemical impedance-based detector. *Biosens. Bioelectron.* **2021**, *171*, 112709. [CrossRef] [PubMed]

31. Yang, L.; Li, Y. Detection of viable Salmonella using microelectrode-based capacitance measurement coupled with immunomagnetic separation. *J. Microbiol. Methods* **2006**, *64*, 9–16. [CrossRef] [PubMed]

32. Delcour, A.H. Outer membrane Permeability and Antibiotic Resistance. *Biochim. Biophys. Acta Proteins Proteom.* **2009**, *1794*, 808–816. [CrossRef]

33. Yang, L.; Li, Y.; Griffis, C.L.; Johnson, M.G. Interdigitated microelectrode (IME) impedance sensor for the detection of viable Salmonella typhimurium. *Biosens. Bioelectron.* **2004**, *19*, 1139–1147. [CrossRef]

34. Bonanni, A.; Fernandez-Cuesta, I.; Borrise, X.; Perez-Murano, F.; Alegret, S.; del Valle, M. DNA hybridization detection by electrochemical impedance spectroscopy using interdigitated nanoelectrodes. *Microchim. Acta* **2010**, *170*, 275–281. [CrossRef]

35. Ciminelli, C.; Conteduca, D.; Cito, M.; Dell'Olio, F.; Krauss, T.F.; Armenise, M.N. New optoelectronic device for study antimicrobial resistance. In Proceedings of the 1st International Conference on Dielectric Photonic Devices and Systems Beyond Visible, Bari, Italy, 1–2 October 2018.

36. Petrovszki, D.; Valkai, S.; Gora, E.; Tanner, M.; Bányai, A.; Fürjes, P.; Dér, A. An integrated electro-optical biosensor system for rapid, low-cost detection of bacteria. *Microelectron. Eng.* **2021**, *239*, 111523. [CrossRef]

37. Cardile, P.; Franzo, G.; Lo Savio, R.; Galli, M.; Krauss, T.F.; Priolo, F.; O' Faolain, L. Electrical conduction and optical properties of doped silicon-on-insulator photonic crystals. *Appl. Phys. Lett.* **2011**, *98*, 203506. [CrossRef]

38. Magnusson, R.; Wang, S.S. New principle for optical filters. *Appl. Phys. Lett.* **1992**, *61*, 1022–1024. [CrossRef]

39. Wang, S.S.; Magnusson, R.J.A.O. Theory and applications of guided-mode resonance filters. *Appl. Opt.* **1993**, *32*, 2606–2613. [CrossRef]

40. Tibuleac, S.; Magnusson, R. Reflection and transmission guided-mode resonance filters. *JOSA A* **1997**, *14*, 1617–1626. [CrossRef]

41. Aspnes, D.E.; Studna, A.A. Dielectric functions and optical parameters of Si, Ge, GaP, GaAs, GaSb, InP, InAs, and InSb from 1.5 to 6.0 eV. *Phys. Rev. B* **1983**, *27*, 985. [CrossRef]

42. Conteduca, D.; Barth, I.; Pitruzzello, G.; Reardon, C.P.; Martins, E.R.; Krauss, T.F. Dielectric nanohole array metasurface for high-resolution near-field sensing and imaging. *Nat. Commun.* **2021**, *12*, 1–9. [CrossRef]

43. Hall-Stoodley, L.; Costerton, J.W.; Stoodley, P. Bacterial Biofilms: From the natural environment to infectious diseases. *Nat. Rev.* **2004**, *2*, 95–108. [CrossRef] [PubMed]

44. Varshney, M.; Li, Y. Interdigitated array microelectrodes based impedance biosensors for detection of bacterial cells. *Biosens. Bioelectron.* **2009**, *24*, 2951–2960. [CrossRef] [PubMed]

45. Yang, L.; Ruan, C.; Li, Y. Detection of viable Salmonella typhimurium by impedance measurement of electrode capacitance and medium resistance. *Biosens. Bioelectron.* **2003**, *19*, 495–502. [CrossRef]

46. Subramanian, S.; Tolstaya, E.I.; Winkler, T.E.; Bentley, W.E.; Ghodssi, R. An integrated microsystem for real-time detection and threshold-activated treatment of bacterial biofilms. *Appl. Mat. Interfaces* **2017**, *9*, 31362–31371. [CrossRef]

47. Ramos, A.; Morgan, H.; Green, N.G.; Castellanos, A. Ac electrokinetics: A review of forces in microelectrodes structures. *J. Phys. D Appl. Phys.* **1998**, *31*, 2338–2353. [CrossRef]

48. Berdat, D.; Rodríguez, A.C.M.; Herrera, F.; Gijs, M.A. Label-free detection of DNA with interdigitated micro-electrodes in a fluidic cell. *Lab a Chip* **2008**, *8*, 302–308. [CrossRef]

49. Schmid, J.H.; Sinclair, W.; García, J.; Janz, S.; Lapointe, J.; Poitras, D.; Li, Y.; Mischki, T.; Lopinski, G.; Chepen, P.; et al. Silicon-on-insulator guided mode resonant grating for evanescent field molecular sensing. *Opt. Express* **2009**, *17*, 18371–18380. [CrossRef]

50. Soref, R.; Bennett, B. Electrooptical effects in silicon. *IEEE J. Quantum Electron.* **1987**, *23*, 123–129. [CrossRef]

51. Liu, P.Y.; Chin, L.K.; Ser, W.; Ayi, T.C.; Yap, P.H.; Bouroina, T.; Leprince-Wang, Y. Real-Time measurement of single bacterium's refractive index using optofluidic immersion refractometry. *Procedia Eng.* **2014**, *87*, 356–359. [CrossRef]

52. Kim, T.; Kang, J.; Lee, J.H.; Yoon, J. Influence of attached bacteria and biofilm on double-layer capacitance during biofilm monitoring by electrochemical impedance spectroscopy. *Water Res.* **2011**, *45*, 4615–4622. [CrossRef] [PubMed]

53. Oberlander, J.; Jildeh, Z.B.; Kirchner, P.; Wendeler, L.; Bromm, A.; Iken, H.; Wagner, P.; Keusgen, M.; Schoning, M.J. Study of interdigitated electrode arrays using experiments and finite element models for the evaluation of sterilization processes. *Sensors* **2015**, *15*, 26115–26127. [CrossRef] [PubMed]

54. Jones, T.B.; Nenadic, N.G. *Electromechanics and MEMS*; Cambridge University Press: Cambridge, UK, 2013.

55. Asami, K. Characterization of heterogeneous systems by dielectric spectroscopy. *Prog. Polym. Sci.* **2002**, *27*, 1617–1659. [CrossRef]

56. Bai, W.; Zhao, K.S.; Asami, K. Dielectric properties of *E.coli* cell as simulated by three-shell spheroidal model. *Biophys. Chem.* **2006**, *122*, 136–142. [CrossRef]

57. Malaquin, L.; Vieu, C.; Martinez, C.; Steck, B.; Carcenac, F. Interdigitated nanoelectrodes for nanoparticle detection. *Nanotechnology* **2005**, *16*, S240. [CrossRef]

MDPI

St. Alban-Anlage 66

4052 Basel

Switzerland

Tel. +41 61 683 77 34

Fax +41 61 302 89 18

www.mdpi.com

Biosensors Editorial Office

E-mail: biosensors@mdpi.com

www.mdpi.com/journal/biosensors